Starting Electronics Construction

Techniques, Equipment and Projects

0750 667 362

D1148450

Starting Electronics Construction

Techniques, Equipment and Projects

Keith Brindley

AMSTERDAM • BOSTON • HEIDELBERG • LONDON • NEW YORK •OXFORD
PARIS • SAN DIEGO • SAN FRANCISCO • SINGAPORE • SYDNEY • TOKYO

Newnes is an imprint of Elsevier

ELSEVIER

Newnes

Newnes
An imprint of Elsevier
Linacre House, Jordan Hill, Oxford OX2 8DP
30 Corporate Drive, Burlington, MA 01803, United States.
First edition 2005

British Library Cataloguing in Publication Data
A catalogue record for this book is available from the British
Library

ISBN 0 7506 67362

For information on all Newnes publications
visit our website at www.newnespress.com

 Typeset by Co-publications, Loughborough

Printed and bound by MP Books Ltd, Bodmin, Cornwall

Working together to grow
libraries in developing countries

www.elsevier.com | www.bookaid.org | www.sabre.org

ELSEVIER   BOOK AID International   Sabre Foundation

Contents

Preface

This book is all about how to make electronics devices, from a constructional perspective. It is aimed at relative newcomers to electronics construction — that is, not the knowledgeable and trained electronics engineer, but those of us who want to find out about how to actually make up the circuits and projects that have already been designed.

It matters not whether the circuits you want to construct have been designed by others — perhaps they are presented in a magazine or a book as projects to build — or by yourself. It matters not whether you actually understand how that circuit works, or whether you haven't the foggiest idea of even its first principles — and maybe you don't even care! It matters not whether the intricate workings of a resistor; transistor; transformer; capacitor; or the latest wonder component is like Greek to double Dutch ears. There are other books that you can read which can give you all that understanding — and I'm bound to suggest my own book *Starting Electronics* as the ideal companion reference to this one, if that's the sort of understanding you are also looking for.

What *does* matter, on the other hand, is that *this* book (at least, that's my intention) shows everyone how to take an existing circuit and transform it into a complete and working device. This book starts from only the most limited knowledge of electronics processes, assuming very little — hopefully nothing at all, and introduces you to *all* the required concepts; the tools; the components; and the processes involved in turning a circuit into a finished electronic device. In fact, my intention in writing it is not only to make the journey from circuit to device as easy as possible to traverse, but to make it fun at the same time. I hope I've succeeded.

Running to keep up...

In electronics, as with most fast-moving technologies, there's always the risk that a book such as this one can become out-of-date very quickly. The way electronics moves these days, it's not unfair to say that a book like this is only up-to-date for as long as the ink is wet. So, at this point, I'd like to introduce readers to a central website:

http://keithbrindley.members.beeb.net

where I will try to keep relevant information about the methods and processes covered here up-to-date. The website also features relevant links to manufacturers, component stockists, and other links as I see fit. Projects (in Chapter 8) all feature printed circuit board copper foil track patterns, and there are links on the website to allow you to download these in electronic formats too. Indeed, I'd go as far as suggesting that if you intend constructing any of the projects here you should look on the website to see if there are any alterations or other suggestions regarding the projects before you start. Any corrections or alterations to the copper foil track patterns (or anything else to do with the projects) will be featured on the website as soon as I can get them there.

As you make it...

Electronics in general is a fascinating area. However, getting 'into it' in the first place is sometimes daunting. It is my aim that *Starting Electronics Construction* gives as many people as possible the means whereby they can do just that. In the meanwhile, if I can offer anyone a suggestion if they intend to try, it is this: make it fun!

That way, you'll learn so much more than you ever could from stuffy textbooks. I hope that by reading this book and getting into practical electronics construction that you do have fun.

Keith Brindley

1

Introduction

This book is a guide to anyone who wants to build an electronic device. It assumes very little prior knowledge, and so introduces you to the various practices that you need to follow — whether you want to build something for pleasure, or for education, or even for profit. Also, this book takes the stance that it matters little what the particular electronic product is you wish to build — the most important fact is that the practices you follow are similar (if not identical) whatever the electronic device happens to be.

Electronics is in part a theoretical entity, and you could go to textbooks or to magazines, schools, colleges or universities, or elsewhere to find and learn that theory. However, it is also in a very large part a practical entity; a fact that — particularly in recent years — seems almost to have been overlooked. The many students of electronics at universities who have little or no prior experience with the handling of components or the building of circuits attest to this.

So, although what you read on the pages of this book is theory, it is theory about the practical aspects of building electronic devices. As such, included in this book are (hopefully most of, possibly all) the practical things you need to know about and be able to do before you can make an electronic device.

What this book is *not*, on the other hand, is a book about how electronics circuits *work* (for that, see this book's sister book: *Starting Electronics*, which is written also by me). What you have in front of you now is a book only about how to *construct* electronic devices. I hope you enjoy it, and gain some knowledge about how to do just that.

What do you need to know about making an electronic device?

This section is intended as a summary of the whole book. It would be easy to miss this out altogether, but my feeling in writing it is that you need to see the *big picture* before you go on to look at things in more detail.

And the big picture in terms of building electronics devices is simply a quick overview of what an electronic device is, and also what parts of it should we understand in order that we can build it.

So, what is an electronic device?

We can summarize an electronic device (an example of which is given in Figure 1.1) as below.

An electronic device invariably has three main features:

NUMBER 1 Electronic components. An electronic device is anything that incorporates electronic components of any description.

NUMBER 2 A printed circuit board. The electronic components which make up an electronic device are mounted upon a *printed circuit board* — (often called just a *PCB*), which is a means of holding all the components securely while at the same time making the necessary electrical connections between them all.

NUMBER 3 A housing. The printed circuit board, together with all the electronic components that it holds, will be housed in a box, or a module, or a case. In this way the electronic components are protected from damage.

So, what's next?

Given that any electronic device you might choose to build has these three main features, it follows that you need to know as much as you can

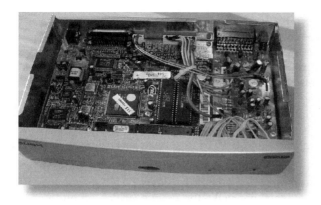

FIGURE 1.1   An example of an electronic device – note the electronic components, the printed circuit board, and the housing

about the features. And we'll look at them now, in brief, before devoting the rest of the book to that purpose.

Electronic components

There are many categories of electronic components, and each category is divided often into several sub-categories, and from there into a multitude of sub-sub-categories. You don't need to know about them *all* by any means, but you do need to know about the most common categories, and their most common sub-categories. You need to know how to handle these components so that you don't damage them, and so you need to know what damage *can* be done to them.

Some components are easily damaged just by holding them a little too roughly; some components are damaged by heat; and certainly many components are damaged if you insert them into a circuit incorrectly (the wrong way round, say, or too forcefully, thereby breaking their connection pins).

You also need to know about which tools you need to handle them and how to use these tools correctly.

Printed circuit boards

Making the printed circuit board onto which the electronic components of a device are mounted is not a trivial job. Indeed, the work involved in designing it, making it, then attaching the components to it represents a major proportion of the overall task. However, making a printed circuit board is not a job which should be rushed, for a poorly made printed circuit board is a recipe for disaster.

The printed circuit board is used for two main purposes: to support the electronics component; and to connect between the electronic components so that the circuit they are to be connected into can be formed. As a result, a printed circuit board that cannot support the components adequately, or makes incorrect connections, will prevent the electronic device from functioning correctly. So, you need to know about all aspects of printed circuit boards: how to design them; how to make them; and how to mount the components. You need to know what tools to use to do these tasks, and how to use the tools correctly.

I simply cannot stress enough how important the printed circuit board is. If the circuit itself is the brain of an electronic device — that monitors, controls and sends the signals that make the device work — then the printed circuit board is the heart that keeps the whole device going in the first place.

Housings

Once the printed circuit board is complete, with all electronic components in place, you have to consider how to house it. Housings are used to protect the components inside, as well as their printed circuit board. Housings also usually give a little extra space to allow batteries to be held, when needed, as well as providing mounting space for some components and controls not mounted on the board. They also protect users from potentially harmful voltages that may be present in the circuit — mains voltages, say. Labelling on a housing's facing panels can indicate what various control adjustments may be made by the user. As such, a housing is the interface — technically, visually, and aesthetically — with the outside world.

In short, if the circuit is the brain, and the printed circuit board is the heart, then the housing is the body, the skeleton, and the skin of an electronic device.

Housings done right can make a product a winner; done wrong the product is a duffer.

What have we learned?

To summarize this chapter we only need to remember one thing — an electronic device comprises three main parts. These main parts — the electronic components, the printed circuit board, and the housing — are highly dependent on each other. Designers and makers of *good* electronic devices understand this dependence, and use the main parts to the best advantage of their devices.

2

Tools

As with any practical technology, having the right tools for the task is important. Indeed, some of the tools are not just important — they are critical! You simply cannot do electronics construction without them. So, first off in this chapter is a hit parade of the tools you need, followed by another of the tools you should seriously consider (although you *could* do without them at a push!).

Tools you absolutely, positively, indisputably, and without a doubt need:

SOLDERING IRON — to solder components into place. The connecting lead of each component must be soldered into place on the printed circuit board. A specialized soldering iron must be used — one that is neither too big nor too powerful to harm either the component or the circuit board

MINIATURE SIDE-CUTTERS — to cut wire or component leads. While constructing an electronic device, there will be many wires to cut to size. Each component mounted on the printed circuit board needs its leads trimming close to the board after soldering

MINIATURE LONG-NOSED PLIERS — to form leads. Component leads are formed to the correct shape for mounting on a printed circuit board with small, long-nosed pliers. Larger components and many controls are attached with nuts and bolts, so a larger pair of long-nosed pliers is also needed

CONTINUED...

...CONTINUED

WIRE-STRIPPERS — to strip insulation from wires. Insulation must be removed from wires prior to soldering, and gripping the wire in your teeth before pulling the insulation off isn't the best method. A good pair of wire-strippers solves the problem and keeps your dental bills as low as possible

SOLDER REMOVAL TOOLS — to err... remove solder. At some point or other, you will need to remove solder from a soldered joint, usually to take out a component you have accidentally put in the wrong place, or to replace a faulty component. Some type of solder-removal tool is required, allowing you to extract the component's leads

DRILLS AND BITS — to err... drill holes. Every component lead goes through a hole in the printed circuit board. Each hole must be drilled. Components not mounted on the printed circuit board are usually mounted on the housing, usually with nuts and bolts, and holes for these must be drilled in the housing

SAFETY GOGGLES — to protect your eyes. Some of the processes involved in electronics construction may create potential hazards. You need to protect yourself! Goggles should be used when undertaking any process likely to create flying debris, such as drilling or using chemicals.

Tools you don't exactly *need* — but which make some jobs easier:

PLIERS — ordinary hand pliers

FILES — a small set of needle files, and a selection of larger files

PUNCHES — for making some larger holes in housings

VICE — to hold housings while working

A STAND — to hold printed circuit boards.

The Tools

We'll now take a closer look at the most important of those tools listed, and note any particular attributes they should have.

Soldering irons

Soldering in electronics does not have any particular requirements, other than the soldering iron should not be too powerful. On the face of it, this might sound odd to a newcomer given that the process of soldering requires solder to be melted thus forming a soldered joint, but it's quite logical when you consider exactly *what* you are soldering — electronic components. The problem is that electronic components can be damaged by too much heat — and the more powerful the soldering iron, the greater its capacity to deliver heat.

Soldering iron power is measured in watts, and a reasonable wattage for soldering electronic components is somewhere between 12 and 25 watts. Prices vary between models, but you should expect to pay between about £10 and £20 for a basic soldering iron, although more expensive soldering irons (with extra features) exist.

A typical soldering iron is shown in Figure 2.1, where you can see its main parts — a handle to hold it, a bit (the part that heats the joint and the solder), and a heating element (the part that heats the bit). Also shown in Figure 2.1 is a mains lead, although not all soldering irons require mains electricity to function, as we'll see soon.

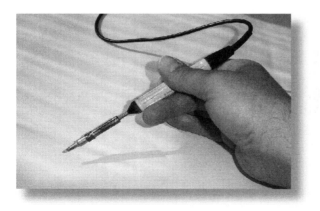

FIGURE 2.1   A typical soldering iron – the main parts of which are (from left to right): the bit; the heating element; the handle. This particular soldering iron is mains-powered, so you can also see the mains lead exiting the handle at the right.

Soldering iron types

There are several variants of soldering irons, in terms of heating methods, listed below.

Mains-powered irons — the most common. The advantage of these is that they work wherever a main outlet is available. The disadvantage is that they do have a mains lead, which some people find a little fiddly in use. Basic soldering irons of this type have no temperature control, and merely get as hot as they can. Soldering irons with thermostatic controls are available, and even complete soldering iron/base station combinations are common (see later). Your budget defines what you buy.

Rechargeable irons — with base station. Rechargeable irons have internal batteries that are used to heat the element and bit. They have a base station that is mains-powered, into which the soldering iron fits when not in use. Connections in the base station and soldering iron mate, and so the batteries are recharged whenever the iron is in the base station (and the base station is powered, of course).

The main advantage of rechargeable soldering irons is their lack of mains lead in use. The main disadvantage is that the batteries can't power the element for longer than just a few minutes at a time. By replacing the soldering iron into the base station after each joint is soldered though (so that its batteries recharge partially), this can be extended somewhat. Figure 2.2 shows a rechargeable soldering iron with its base station.

Figure 2.2 Rechargeable soldering iron in use. To conserve batteries, the element only heats up when the button on the handle is pressed – heating rapidly to soldering temperature within seconds. In the background is the base station which recharges the soldering iron's batteries.

Gas-powered irons — the ultimate in portability. Fitted with small refillable charges of butane gas which last up to around 30 minutes of constant use, these soldering irons require no electrical power. Figure 2.3 shows a gas soldering iron in use.

Priced between around £20 to £40 or so, these soldering irons are useful for certain awkward soldering jobs where electrical power is not available. They tend to be quite powerful, particularly the more expensive models, though with care can be used to good effect in electronics soldering.

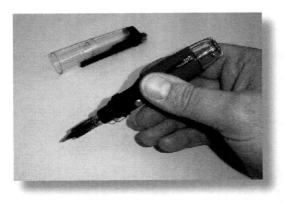

FIGURE 2.3 A gas-powered soldering iron. Small and portable, these can be recharged from lighter fuel butane gas containers.

Low-voltage irons — for battery-powered applications. Occasionally, when working on a car, say, there are applications where a low-voltage soldering iron might be useful. Some 12 V-powered soldering irons are available (an example is shown in Figure 2.4), usually having battery clips attached to the power lead.

Soldering iron stations
While we're on the subject of soldering iron types, it's worth taking a look at soldering iron stations. These are complete units, comprising a soldering iron, a stand to hold the iron when not in use, and a controlling circuit of some description. A typical soldering iron station is shown in Figure 2.5. The controlling circuit typically controls the power to the soldering iron element, so that the bit stays at a constant, definable temperature. While such stations are expensive, ranging from around £30 to over £100, they are excellent tools if you do a considerable amount of soldering.

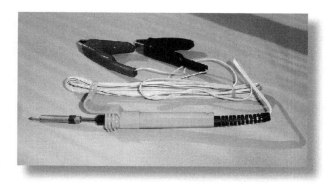

FIGURE 2.4   A 12 V-powered soldering iron. The battery clips on the lead attach to a car's battery for (relatively) portable power.

FIGURE 2.5   A soldering iron station, showing the iron itself in a stand, and the control knob to set bit temperature (Antex Electronics).

Soldering iron bits

Alongside the iron itself, you need to think about the bit. The bit is the part of the iron that comes into contact with the parts to be joined, so is a fairly critical part of the process.

A typical range of bit shapes is shown in Figure 2.6. Usually, for soldering electronic component joints an angled bit of about 2.4 mm diameter is ideal.

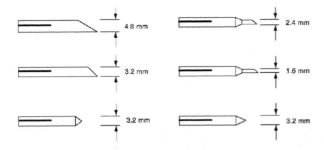

FIGURE 2.6   Typical sizes of soldering iron bits.

How the bit fits onto the soldering iron element is of concern, for two reasons:

- it determines how much of the heat generated by the iron's element passes to the bit's tip
  — and hence to the joint to be soldered

- it determines how easy it is to remove the bit
  — important if you are to solder different types of joints (for example, a bit with a smaller diameter would be used for finer joints), or for when you need to replace an old and worn bit.

In general, bits that slide over a soldering iron's element are most efficient, allowing much of the heat generated by the element to pass to the bit's tip. Figure 2.7 shows such a bit, fitted to its element. Also, this bit type is the easiest to remove.

Bits that fit inside an element, or screw onto it, are probably best avoided, if only because the very nature of the soldering process means that metal bits often corrode — even slightly — due to heat, and this may mean that the bit becomes hard to remove. Excessive pulling or twisting of the bit to remove it may then damage the element.

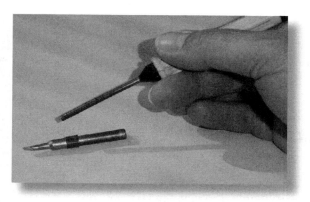

FIGURE 2.7   A soldering iron whose bit slides over the element. The bit is easily removed and/or replaced, and provides an efficient heat transfer between the element and the bit's tip.

Some soldering irons use a bit type that also comprises the element. The bit/element combination attaches to the soldering iron by two prongs that slide into the soldering iron body. Element and bit tip being unified in this way, there is a highly efficient heat transfer between element and tip, and this design has the great advantage that changing the bit effectively replaces the element in one operation. One of the best examples is the rechargeable soldering iron bit tip shown in Figure 2.8.

FIGURE 2.8   The unified element/bit of the rechargeable soldering iron shown in Figure 2.2 fits to the soldering iron's body by pushing in the two prongs.

Soldering iron accessories

Soldering iron stand

There's a few accessories for soldering irons that are worth buying. First
— and most important — of these is a soldering iron stand (a typical
stand is shown in Figure 2.9), which holds an iron while it's not in use.
Soldering irons, of course, by their very nature are hot, so having a stand
to put one in when not actually using it adds up to quite a safety feature.
The stand in Figure 2.9 is around a fiver — so shouldn't break the bank.

FIGURE 2.9   A soldering iron stand. This particular stand features a sponge.

Sponge

Talking of a soldering iron stand with a sponge, a sponge is actually a
very handy accessory to have. Even if you have to resort to — as I have
done on occasion — using a simple kitchen sponge, dampened with
water, stood on a tea-plate, then so be it!

The £5 soldering iron stand of Figure 2.9, featuring an integrated sponge
is a great way to combine the two accessories — not essential, just neat.

You use a dampened sponge to wipe your hot soldering iron bit tip on,
immediately prior to tinning it (see page 117) and using it to solder
joints. As you'll see later, a clean soldering iron bit tip is one of the
prerequisites of good soldering.

Put another way: if your soldering iron bit tip is not clean, don't expect to
make good soldered joints, and *do* expect your circuit not to work!

Tip tinning and cleaning block

Another way to keep your soldering iron bit's tip clean is to use a tip tinning and cleaning block. It's a thin disc-shaped block in a metal container that can be carried in a pocket, or mounted on a suitable surface close to the soldering iron. In use, you simply wipe the hot soldering iron bit's tip over the block's surface, simultaneously cleaning and tinning the tip, as shown in Figure 2.10. It's convenient, it's handy, it's cheap (around £4), and it saves having a wet sponge on your worktop.

FIGURE 2.10   Using a tip tinning and cleaning block to err... clean and tin a hot soldering iron bit tip.

Abrasive fibreglass pencil

While we're on the subject of cleanliness (which we'll return to often throughout the book), the other components that go to make up a soldered joint have to be clean — just as much as the soldering iron bit tip. Fortunately, there are products that can help here too. The copper surfaces of the track on a printed circuit board should be clean and, as they oxidize when left for more than just a few hours (and copper oxide can prevent a good joint from being made), they should be cleaned more-or-less immediately prior to soldering the joints.

A good tool for this job is a fibreglass abrasive pencil. It's a propelling-pencil type of design, with an internal fibreglass stick instead of pencil lead — see Figure 2.11. In use, you rub the end of the fibreglass stick on the copper track a few times, removing any dirt, grease, or pre-finishes, leaving clean, shiny copper, ready for soldering.

FIGURE 2.11   An abrasive fibreglass pencil in use — note the residue, which is of fibreglass origins.

After cleaning the copper surface, care should be taken when removing the residual dust created in the process — it *is* fibreglass-based after all, so may irritate skin, eyes, or internal organs.

Abrasive scrubbing block
An alternative to the fibreglass pencil is an abrasive scrubbing block, which performs the same function — scrub the copper track prior to soldering with the block, in an action reminiscent of using an eraser to rub out pencil marks. Figure 2.12 shows such a scrubbing block.

FIGURE 2.12   An abrasive scrubbing block.

Component leads — which form a third critical part of a soldered joint on a printed circuit board need to be clean every bit as much as the copper track and the soldering iron bit's tip. Most components are supplied with tinned leads (ie, they are coated with molten solder then cooled), fortunately, which means that a simple reheating to soldering temperature re-melts the solder which helps the process of soldering the joint. Even so, excess dirt — particularly grease — on a lead may still prevent a good joint being made, so if you suspect that to be the case you should clean the lead prior to soldering. It's much easier to do this before soldering, than trying to correct the effects of it afterwards.

Chemical solvent cleaners

Chemical cleaners such as isopropyl alcohol may be useful for cleaning printed circuit board tracks or component leads prior to soldering, as they dissolve grease, dirt, and some pre-finishes which may be present. They can be purchased in liquid form, although aerosol or pen applicators are often much more convenient.

Heat shunt

The final soldering accessory you might want to consider in your armoury is a heat shunt. Soldering — by its very nature — requires heat, but too much heat can damage some electronic components. Some people therefore like to use a heat shunt between the soldered joint and the component body to help ensure the heat at the joint is shunted away from the component itself.

The simplest form of heat shunt is a pair of locking tweezers which are clamped onto the component lead. As the joint is heated, heat is shunted into the tweezers rather than passing down the component lead into the component body.

Figure 2.13 shows such a pair of tweezers.

> **Take Note:**
>
> Having said all that, I am duty-bound to say that I have never owned a heat shunt, I have never even used a heat shunt, and I don't think I will ever use a heat shunt. All but a very few electronic components are fairly heat-resilient after all, and as long as the soldered joint is performed rapidly and without leaving the soldering iron in place on the joint too long, then excessive heat should not be a problem.

FIGURE 2.13   A heat shunt.

Side-cutters

After your soldering iron, the single most important tool in your toolbox is a decent-to-good quality pair of miniature side-cutters. As side-cutters *are* so important, it's best not to skimp on this tool — buy the best you can afford (decent quality side-cutters are available at around £10, while *good* quality side-cutters are priced between about £20 and £80...).

Before you baulk at the price, remember that the side-cutters you buy will be used every time you solder a printed circuit board, to trim the component leads after soldering joints. They'll be used every time you cut wires to length, for insertion into a printed circuit. They'll be used every time you trim your toenails. They should *not* be used for cutting thick wires or metal (or pruning the roses, or cutting toenails, if you hadn't already realized I wasn't being serious).

And finally, just to labour the point, a decent pair of side-cutters will last a lifetime in electronics, but a cheap pair will probably last till you need them.

Figure 2.14 shows a pair of side-cutters that have lasted a lifetime (mine). They were expensive (costing around £50 in 1976), and at the time I was young and foolish and thought it was a little over-the-top, but now that I've matured and I'm 21 again (err... yes, well...) I certainly realize their worth. In all seriousness though, the money you spend on this tool is worth it in the long run.

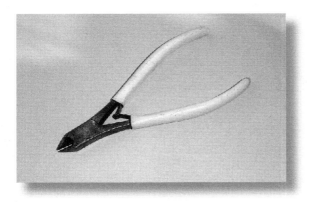

FIGURE 2.14   A good quality pair of miniature side-cutters. After your soldering iron, it's the tool you'll use the most in electronics.

Long-nosed pliers

Like side-cutters, a miniature pair of long-nosed pliers is an essential tool for the builder of electronics circuits. Like side-cutters also, it's pretty important to have a decent-to-good quality pair. And, if you haven't already guessed it, like side-cutters too, they may be expensive — decent quality long-nosed pliers will cost around £10, while a good quality pair will be priced between around £20 and £80). Again, get the best you can afford — they will be worth it. Figure 2.15 shows my long-nosed pliers (which I bought along with my side-cutters in 1976, for around £50). They're still working perfectly, which is more than I can say for myself.

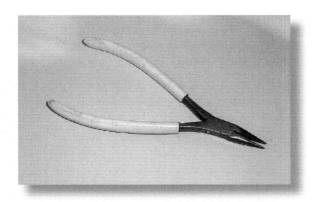

FIGURE 2.15   Miniature long-nosed pliers are extremely useful tools to keep in your electronics toolbox.

Long-nosed pliers are very useful for forming component leads prior to mounting the component in the printed circuit board. They should not be used for anything heavier than this — and should definitely *not* be used for tightening nuts onto large bolts, or working on the car. They are precision instruments which require a little forethought in handling.

Wire-strippers

Printed circuit boards are not usually complete entities, and they need to be connected to the outside world. The easiest and most popular way to do that is with interconnecting wires, which are normally insulation-covered to prevent short circuit with their neighbours. Each wire has to be soldered to the printed circuit board, so it's necessary to remove a piece of insulation at the wire end, leaving a short length of uninsulated wire that can be soldered. It's perfectly possible to strip a short piece of insulation away by putting the end of the wire into your mouth between your teeth, biting on the wire while pulling it back out of your mouth. Indeed, I used to do this very thing before I wore a hole in my front teeth and suffered many expensive dental bills. Now I use a pair of wire-strippers, and I wholeheartedly recommend you do the same.

Wire-strippers work in exactly the same way that front teeth do, in that you insert the wire into the wire-strippers jaws (and yes, they are called jaws) whereupon you squeeze the wire-strippers' handles together so that the wire-strippers' teeth catch the wire between them. Further squeezing of the handles forces the teeth into the insulation and simultaneously pulls the jaws back, stripping the insulation from the wire. Figure 2.16 shows a pair of wire-strippers in action.

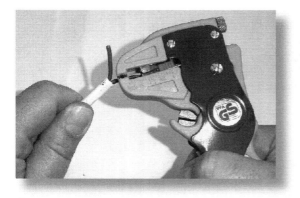

FᴵGᴜʀᴇ 2.16   Using a pair of wire-strippers to strip the insulation from the end of a wire, prior to soldering to a printed circuit board.

Such wire-strippers aren't too expensive (around a tenner), although you can get much more expensive ones. As with the case of wire-cutters and long-nosed pliers, get the best you can afford. The best wire-strippers have replaceable jaws (unlike me) so can be always maintained with a good set of teeth. The cheaper ones are just throw-away tools (I've heard *that* said about me before, as well...).

Solder removal tools

If you solder electronic components into printed circuit boards, then sooner or later you will need to desolder them to remove them from printed circuit boards. It goes without saying that you will at some stage put a component in the wrong place and need to take it out, or you will find that you have a faulty component and need to replace it.

There are two tools that can help you do this and your selection of which you buy is merely a matter of choice. The cheapest is desoldering braid — lengths of copper braided wires that attract molten solder away from a soldered joint, simply because solder flows to cover the copper wire surfaces in the braid. Figure 2.17 shows desoldering braid in use — the braid is placed over the joint and a hot soldering iron bit presses on top. As the heat from the iron bit melts the solder it is attracted onto the braid away from the joint. Once the component lead becomes sufficiently free of solder to be removed, the braid and the soldering iron bit are removed, and the lead can be extracted from the joint hole.

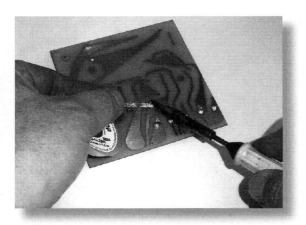

FIGURE 2.17   Desoldering braid in action. The braid is heated by the iron's bit, which in turn heats the solder on the joint to melting point, whereupon the solder is attracted to the copper wires in the braid.

Once the braid end has been used to desolder a joint, it becomes loaded with solder so must be cut off, leaving fresh braid for the next desoldering operation. As such, a length of desoldering braid only has a limited life (depending on how many joints you desolder). Desoldering braid costs £2 or £3 for a metre or so, although you can buy it in longer length reels if you need.

The alternative tool for desoldering purposes is a desoldering pump, commonly called a *solder sucker*. This works something like a bicycle pump in reverse — you operate the pump (which is spring-loaded) to push out air, then apply the pump to molten solder on a joint and release the plunger to literally suck the molten solder away from the joint into the pump where it rapidly hardens. The next operation of the pump forces out the hardened solder and primes the pump ready for the next desoldering operation. Figure 2.18 shows a desoldering pump in position over a joint as the joint is heated to become molten by a soldering iron bit tip. The pump has been primed (ie, its plunger has been pushed down the pump body), and as the solder becomes molten, the button catch on the desoldering pump body is pushed, thereby releasing the spring-loaded plunger which returns to the top of the pump body.

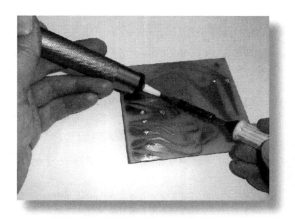

FIGURE 2.18   A desoldering pump in use. As the plunger returns to the top of the pump body, suction it creates sucks the molten solder on the joint inside the pump.

While a little more expensive than desoldering braid, desoldering pumps last considerably longer. Indeed, really the only part of the pump that may be worn by the desoldering process is the pump nozzle, due to the repetitive application of the heat of soldering, which nevertheless will have an operating life of many thousands of operations. Most desoldering pumps have a replaceable nozzle, so even if the nozzle *is* worn, the pump is not necessarily written off. Desoldering pumps are available from as little as £5 to £20.

Drills and bits

Each and every component mounted on a printed circuit board in the conventional way (note I say *conventional way* here: there *are* other methods which don't normally affect the individual electronic circuit builder, most notably the surface mounted component method — see page 36) has its component leads going through holes in the printed circuit board. Thus, the printed circuit board has to be drilled with holes for those very same component leads. The hole size is around 1 mm (give or take a very small percentage — depending on the actual component's leads), which means that your average power drill that you use to drill masonry or steel is not exactly the best tool to use. Power drills are good where the thing being drilled is hard, and the holes don't have to be too accurately positioned.

Printed circuit board, on the other hand, is made from a thin and fairly soft substrate, and the copper track is but a few micrometres thick, while the copper pads you'll be drilling through are not that much bigger than the holes required themselves. So, a drill for the task of drilling printed circuit board has few power requirements, but accuracy is of prime importance.

Small, hand-held, low-voltage drills (complete with mains transformer) are available from most do-it-yourself stores, and these are ideal for hand drilling of printed circuit boards. Figure 2.19 shows such a drill, being used to drill holes in a printed circuit board, prior to mounting and soldering of components.

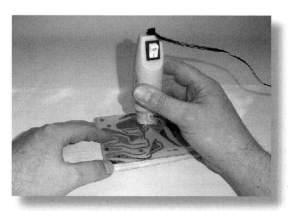

 ▷ **FYI** ◁

Make sure you keep a couple of 1 mm and 1.5 mm drills in your toolbox at all times – it's useful when one drill breaks (which they invariably do at the wrong time) to have another one ready to use. Oh, and make sure all your drills are sharp.

Figure 2.19   Using a hand-held low-voltage drill to drill holes in a printed circuit board.

Later in the book we'll see that larger power drills do actually have a place in the electronics circuit builder's toolbox, for drilling of large holes in housings for example. But you *do* need to have a low-voltage, hand-held drill for the finer tasks like that in Figure 2.19. They are not that expensive — starting at around £10 and going up to about £30. Even the cheapest will do the job effectively. Usually they are sold in kits, complete with a range of drill bits, together with small grinding wheels, polishing wheels and engraving tools. If you buy a kit, complete with a range of drills, then you will probably have all the drill sizes you need. If you buy the drill alone, make sure you also buy drills of 1 mm and 1.5 mm sizes — they are the most common ones you'll need.

Printed circuit board stand
When working on a printed circuit board by hand, it's not always easy to mount and solder components without the board moving along the worksurface as you work. If you have a hot soldering iron in one hand, and solder in the other hand, the last thing you want is your printed circuit board sliding along the bench and falling to the floor. Face it: you could do with a third hand! An alternative is to use a printed circuit board stand, into which your printed circuit board is clamped. The board can be repositioned easily, and even turned over to suit.

A typical stand is shown in Figure 2.20, which is a miniature vice with a lever-operated sucker arrangement on the bottom. The sucker allows the vice to be clamped to a work surface, while the vice holds a printed circuit board safely, leaving your hands free to insert and solder components.

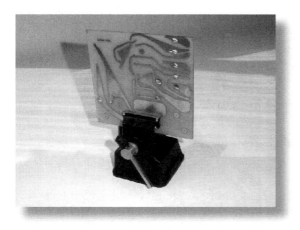

Figure 2.20   A printed circuit board stand.

Many other types of stand are available, ranging from a simple box that holds a printed circuit board, to a multi-armed arrangement complete with magnifying glass and clips to hold other components.

IC insertion tool

As you construct electronic circuits on printed circuit boards, you will come across integrated circuits (ICs — see page 53). Integrated circuits are components that contain many miniaturized components within. They are quite small, but usually have many relatively fragile connection pins (most commonly in two rows — the *dual-in-line* arrangement).

Inserting such integrated circuits into a printed circuit board is a little tricky, as the rows of connection pins are angled somewhat away from the vertical plane, so need to be sprung in as they enter their corresponding holes in the printed circuit board. It's perfectly possible to do this by hand with care, but an IC insertion tool (such as that shown in use in Figure 2.21) is well worth the cost (often less than £1) and makes the task a bit of a doddle.

In use, the integrated circuit is pushed into the tool (which keeps the connection pins at the correct angle and distance), then the tool is pushed down onto the printed circuit board to engage the integrated circuit into its correct position.

FIGURE 2.21   An integrated circuit insertion tool in operation.

IC extraction tool

Getting an integrated circuit back out of a printed circuit board is an equally tricky task. However, there are equally cheap extraction tools — one is shown in use in Figure 2.22. Just hook the tool at each end of the integrated circuit, and pull. Of course, if an integrated circuit is soldered directly into a printed circuit board, you'll need to desolder it first (see earlier). However, the integrated circuit shown in Figure 2.22 has been inserted into a holder. Integrated circuit holders provide a push-fit method of inserting integrated circuits (see Chapter 3).

FIGURE 2.22   Using an extraction tool to pull out an integrated circuit – note the use of an integrated circuit holder, which means the integrated circuit itself is not soldered into the printed circuit board.

Terminal pin pusher

When connections are needed on a printed circuit board, it's common to use push-fit terminal pins (see Chapter 3) in the printed circuit board. It's then easy to solder a connecting lead to the pin (see Chapter 5) as required. However, while it's possible to push these pins into place in a printed circuit board using pliers, say, a good tool to buy is a terminal pin pusher — shown in use in Figure 2.23.

FIGURE 2.23   Terminal pin pusher – just put the pin loosely in place in the printed circuit board, then push home firmly with the pusher.

Spot-face cutter

When using stripboard to build a circuit on, it's inevitable that some of the strips need to be broken — particularly when siting a dual-in-line integrated circuit, which has two rows of parallel pins. In this case, if strips aren't broken between the rows of pins, short circuit between the pins will occur.

Breaking the strip can be done with a sharp craft knife, but this is messy, and a potentially dangerous method. Best to buy a spot-face cutter, which is a tool with a rotating cutting edges. Figure 2.24 shows a common type. Chapter 4 shows how to use one.

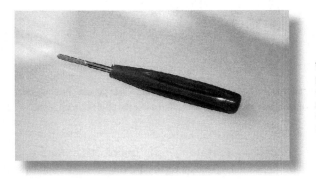

▷ **FYI** ◁

While stripboard is fine for prototype work, I would not condone it for use in a final project. Always use printed circuit board to produce best end results.

FIGURE 2.24   A spot-face cutter, used to break strips in stripboard.

3

Components

There are many types of components used in electronic devices — the main types are listed below. You need to know what each of them are, what their circuit symbols are, how they function, how to identify them, and what their units of measurement and values are.

Component types

- **RESISTORS**

- **CAPACITORS**

- **SEMICONDUCTORS**

- **SWITCHES**

- **INDUCTORS**

- **CONNECTORS**

There are other types of components, but those listed here and described in this chapter are the most important.

Resistors

The resistor is one of the most (if not *the* most) common yet important components used in electronics. Its primary function is to resist the flow of electricity in a defined manner.

To do this it has a resistance, specified in the unit of ohms (named after the physicist Ohm, who did considerable early work into the qualities of resistance) and given the abbreviation Ω. So, a resistor's value is given as a number followed by the ohms symbol eg, 100 Ω. Occasionally, the abbreviation Ω is changed to the abbreviation R — simply because the abbreviation Ω is not always easily produced in word processing applications on some personal computers. Nevertheless, you should know that the R stands for Ω.

Values of resistors range from a fraction of an ohm, through to many millions of ohms. Where thousands of ohms are specified, it is usual to use the unit of kilohms (a kilohm is 1000 ohms), so that say, 2 kΩ is used to specify 2000 ohms), or 470 kΩ specifies 470,000 ohms. Likewise millions of ohms are known as megohms, so that say, 2.2 MΩ is used to specify 2,200,000 ohms, and so on. With kilohms and megohms it has become usual to drop the superfluous Ω, thus values would be written 2.7 k, 470 k, 2.7 M and so on.

On some circuits (and even in some texts or lists), in an effort to decrease the digits and hence size of labels and complexity of the circuit, the ohmic unit is used to indicate the decimal point position — so 8.2 Ω would be written 8R2, 2.7 k would be written 2k7, and 3.3 M would be written 3M3.

Common symbols for a resistor in circuits are shown in Figure 3.1. The correct British standard is the rectangular box shape of Figure 3.1(a), though the zigzagged shape of Figure 3.1(b) is often used, so you must be aware of and be confident with both.

(a)   (b)

FIGURE 3.1   Resistor symbols commonly found in electronic circuits (a) correct British standard (b) zig-zag shape also used.

Apart from a resistor's value — which is by far the most important criterion — there are other important differences which define how well a given resistor works in a given circuit:

- **TOLERANCE — resistors are rarely the exact value specified.**
  Usually resistors are within a small percentage of the specified value — this small percentage is called their tolerance. So, a resistor of value 1000 Ω, which has a tolerance of ±5%, will have a value between 950 Ω and 1050 Ω

- **POWER RATING — as electricity flow causes heat.** If too much heat is generated in a resistor, the resistor may be damaged. Therefore resistors have a power rating (specified in units of watts, and given the abbreviation W) which must not be exceeded. A typical power rating for resistors used in common electronic devices is 0.25 W — in all but the most power hungry of circuits this will usually suffice. In fact, lower power resistors (eg, 0.125 W) will probably do — and are of course smaller

- **STABILITY — the ability of a resistor to keep its value over time.**
  Despite other changes (such as heat, humidity) that can occur around it, it should be able to maintain its resistance value as closely as possible.

Resistor types
There's a few important types of resistor, classified by how they are created.

Carbon resistors
There are actually two main types of carbon resistors, but they look similar if not identical — see Figures 3.2 and 3.3.

FIGURE 3.2   Carbon resistors – illustration.

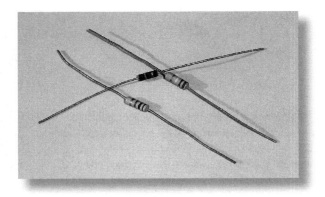

FIGURE 3.3   Photograph of carbon resistors.

Carbon composition resistors are made from a mixture of carbon and clay. Carbon conducts electricity but clay does not, so by mixing the two, resistors are formed. Stability and tolerance of resistors made this way are not particularly good, but they are cheap to make so are very common.

Carbon film resistors are made by moulding a layer of carbon material onto an insulating rod then etching a spiral track through the carbon till the resistance increases. Tolerance and stability are good.

Metal film resistors

These are made in much the same way as carbon film resistors, but tolerance and stability are much better so they are used in circuits which require great accuracy. They are however, more expensive. Figure 3.4 shows illustrations of some metal film resistors, while Figure 3.5 shows a photograph.

FIGURE 3.4   Metal film resistors – illustration.

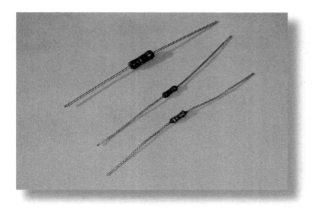

Figure 3.5 Photograph of metal film resistors.

Wirewound resistors

Wire is wound around and along a ceramic former to make these resistors. The thickness of wire and the size of the former define the resistance of the resistor. While being quite expensive, resistors made this way have the advantage of high power rating. Figures 3.6 and 3.7 show illustrations and a photograph of wirewound resistors.

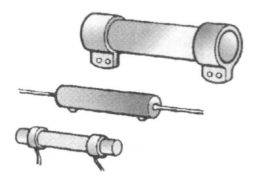

Figure 3.6 Wirewound resistors – illustration.

Variable resistors — potentiometers

Often in an electronic circuit there is a requirement for a control knob of some sort (volume, tone, voltage and so on). Variable resistors (usually called *potentiometers*, or just *pots*) can be used for this purpose. They feature a resistive track usually made from a carbon composition mixture in a circular format, with a rotating wiper arm whose position can be

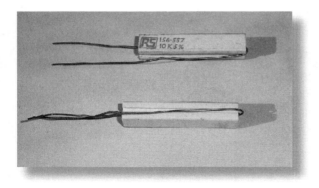

Figure 3.7   Photograph of wirewound resistors.

adjusted. There are two main types — standard potentiometers, featuring a central shaft that can be attached to a circuit's front panel for user-control, and *preset potentiometers* which are mounted on a printed circuit board and are adjusted with a screwdriver-type tool. Their various symbols are shown in Figure 3.8.

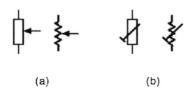

(a)   (b)

Figure 3.8   Symbols of (a) standard and (b) preset potentiometers.

Potentiometers can be of one of two main groups, *logarithmic* (or *log*) and *linear* (or *lin*) — which refers to whether the resistance along the track varies in a linear manner (ie, the resistance varies directly according to the position of the wiper arm) or in a logarithmic manner (ie, the resistance varies logarithmically according to the wiper arm position).

Mostly, linear potentiometers are used. However, in audio circuits in particular — where volume is being adjusted, say — then logarithmic potentiometers are used.

▷ **FYI** ◁

Volume is a logarithmic variant, in so far as a doubling in volume means a 3dB louder sound. So a sound of say 60dB is double the volume of a sound of 57dB. Similarly, a sound of 63dB is double the volume of the 60dB sound. To represent and control this with a potentiometer, the potentiometer must have a logarithmically scaled track. Logarithmic and linear scaled potentiometers can be used and will work in circuits requiring the use of the other type, but the characteristic of the potentiometer control as you turn it will not be correct.

Dual-gang potentiometers are common, particularly in audio circuits where both channels of a stereo amplifier, say, need to be adjusted simultaneously. These are quite literally made of two potentiometers coupled together with a single control that rotates both wipers together.

Most potentiometers are rotary, but sometimes you will see potentiometers that operate in a line. Such *slider potentiometers* work in exactly the same way — with a resistive track and a wiper arm.

Standard potentiometers are relatively large, and the shaft allows for a control knob to be fitted for user control. Switch mechanisms can also be incorporated into the back of the potentiometer body, which means that the control can be used as an on/off switch, say, at the same time as it being a volume control for example. Preset potentiometers are much smaller and get their name because they are typically set before the completed electronic device is first used, then rarely touched or adjusted again. All potentiometers — whether standard or preset — have three terminals: one for each end of the internal resistor, and one for the wiper arm. It's not always necessary to use all three terminals in a circuit though, and often one of the terminals is left unconnected.

Figure 3.9 shows illustrations of various potentiometers, while Figure 3.10 shows a photograph.

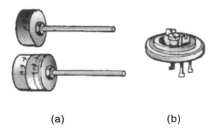

(a)   (b)

FɪɢᴜʀᴇＺ**3.9**   Potentiometers – illustrations: (a) front-panel fixing types – single- (top) and dual-ganged; (b) a preset potentiometer.

Figure 3.10 Photograph of various potentiometers: single-gang, dual-gang, switched single-gang, together with preset potentiometers.

Resistor colour code

Resistance value is occasionally printed numerically on a resistor's body, but normally it is designated in what is called the resistor colour code. The colour code is a collection of coloured bands (normally four bands, although five-band codes exist too) around the resistor body. Reading from left to right, with the first band being closest to the left-hand edge of a resistor, the bands signify:

- **BAND** 1 — the first digit of the resistor's value

- **BAND** 2 — the second digit of the resistor's value

- **BAND** 3 — the figure to multiply the first two digits by

- **BAND** 4 — the resistor's tolerance.

Figure 3.11 shows this graphically. Table 3.1 gives the bands' respective values.

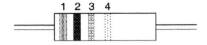

Figure 3.11 A resistor showing the four coloured bands that give its value according to the four-band resistor colour code detailed in Table 3.1.

Table 3.1   The four-band resistor colour code.

Colour	Band 1 1st figure	Band 2 2nd figure	Band 3 3rd figure	Band 4 tolerance
Black	0	0	x1	—
Brown	1	1	x10	1%
Red	2	2	x100	2%
Orange	3	3	x1000	—
Yellow	4	4	x10,000	—
Green	5	5	x100,000	—
Blue	6	6	x1,000,000	—
Violet	7	7	—	—
Grey	8	8	—	—
White	9	9	—	—
Gold	—	—	x0.1	5%
Silver	—	—	x0.01	10%
None	—	—	—	20%

Examples using the resistor colour code

Example 1 — let's say the four bands of a resistor are yellow, violet, orange, silver. The resistor value is 4 (yellow) 7 (violet) multiplied by 1000 (orange), which equals 47,000 ohms (more usually written 47 kΩ), with a tolerance of ±10% (silver).

Example 2 — if the four bands are grey, red, green, gold, then the value is 8 (grey) 2 (red) x100,000 (green), which is 8,200,000 ohms (more usually written 8.2 MΩ), with a tolerance of ±5% (gold)

Preferred values

Initially, when considering the resistor colour code, you may well think that it's a virtually impossible task to remember every single value of resistor together with its respective colour code. Calculating every resistor's value from the colour code under these circumstances will be a laborious task. However, it would be impossible for manufacturers to make, and for stockists to stock, resistors of all possible values — there would simply be too many to deal with effectively. As resistors are only made to certain tolerances anyway, and as circuit designers have a reasonable command over the components they specify for use within their circuits, it has become normal to use resistors of only certain values — these are called *preferred values*. The E12 standard range of preferred value resistances follows the series:

1, 1.2, 1.5, 1.8, 2.2, 2.7, 3.3, 3.9, 4.7, 5.6, 6.8, 8.2

As a result, your job of learning the colour code is made significantly easier — because these twelve values are the only ones you need to know. Any resistors you might use of higher values follow in this very same series merely by having a particular multiplying coloured band (eg, 1.8 x 1000 = 1k8, 4.7 x 1,000,000 = 4M7). After only a short while using resistors, you should find that you can almost instantly calculate any new resistor's value, and you won't even need to calculate resistor values you use most regularly — you'll know them!

Note though, that there are other preferred value ranges of resistors — notably the E24 range, which adds an extra twelve values above the E12 range (spaced in-between the E12 range values). You will rarely, if ever, encounter them however, and the twelve values of the E12 range will most likely be the only ones you need to know.

Special resistors

There are several other types of resistors you may encounter in your electronic travels. Some of these are covered elsewhere in the book, but for completeness the main ones are listed here:

Surface mount resistors — there has always been a trend to reduce electronic devices in size. The latest step in this trend is to use miniature components that are soldered directly to a printed circuit board's copper track ie, they are mounted on the board surface (see Figure 3.12), rather than with component leads that go through the board. As such, surface mount resistors are (by their nature) extremely small, so they are not as suited to hand assembly and soldering as conventional resistors with leads. However you may find yourself working with them occasionally.

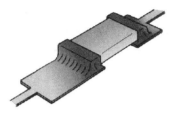

FIGURE 3.12   A typical surface mount resistor – colour coding or lettered markings may be used for component identification (but may not be...). Note that this is illustrated rather larger than full size – actual size is around 2 mm wide!

Thermistors — a thermistor is a resistor whose value changes according to heat. These are useful in applications that detect heat such as electronic thermometers. They vary in size and shape according to construction (they are actually made using semiconductor materials), and typical devices are shown in Figure 3.13. The symbol for a thermistor is shown in Figure 3.14.

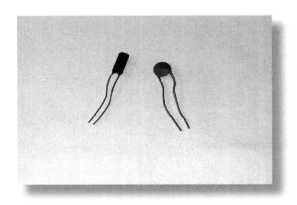

FIGURE 3.13   Typical thermistors.

FIGURE 3.14   Thermistor circuit symbol.

Light dependent resistor — like the thermistor, the light dependent resistor (or LDR) is a resistor made from semiconductive materials (typically cadmium sulphide). Instead of its resistance varying with heat however, the light dependent resistor's resistance varies with applied light. Figure 3.15 shows a typical light dependent resistor — a device known as the ORP 12. The circuit symbol for a light dependent resistor is shown in Figure 3.16.

Fɪɢᴜʀᴇ 3.15   The ORP 12 light dependent resistor.

Fɪɢᴜʀᴇ 3.16   Circuit symbol of a light dependent resistor.

Capacitors

The second most common component used in electronic devices is probably the capacitor. The basic construction of any capacitor is of two metal (therefore conducting) plates, separated by an insulating layer which is called the *dielectric*. The size of the metal plates, together with the type and thickness of the dielectric, defines the capacitance of the component. Capacitance is measured in units of *farads*, named after the scientist Faraday, and given the abbreviation F. However, the farad is an extremely large unit, and capacitors used in modern electronic circuits are only minute fractions (millionths, thousand millionths, and million millionths) of a farad. One millionth of a farad is known as a microfarad (given the abbreviation µF — µ is the Greek letter *mu*, which is often difficult to produce on some computer systems, so the letter *u* is occasionally used instead), one thousand millionth of a farad is a nanofarad (nF), and one million millionth is a picofarad (pF).

It's useful to be able to switch between the three main sizes of capacitance units, as for example one thousand picofarad is the same as one nanofarad, and one thousand nanofarads is one microfarad. Table 3.2 shows the relationships between the three common units. You really need to be able to switch from one to the other easily, so it's worth getting to grips with it fairly early on in your electronics career. Fortunately, you'll get plenty of practice the moment you start to handle capacitors...

Table 3.2 Relationships between the three main capacitance units in modern electronics.

	microfarad (µF)	nanofarad (nF)	picofarad (pF)
microfarad (µF)	1	1000	1,000,000
nanofarad (nF)	0.001	1	1000
picofarad (pF)	0.000001	0.001	1

Electrolytic vs non-electrolytic

There are two main categories of capacitor — *electrolytic* and *non-electrolytic*. While this sounds tremendously complicated and worrying, all it actually means in practice is that electrolytic capacitors are polarized and must be inserted into a circuit the right way round. They will be marked with a positive and a negative terminal, and must be inserted so that the positive terminal is at a positive potential with respect to the negative terminal. (Note that neither the positive nor the negative terminal need to be at an *actual* positive or negative voltage respectively — they just have to be *more* positive or *more* negative than the other respective terminal!) If they are inserted the wrong way round the insulating layer forming the dielectric can be damaged hence won't work correctly, and the capacitor may even become a short circuit hence causing possible damage to the circuit in which it is installed. Non-electrolytic capacitors can be inserted into circuit either way round. Figure 3.17 shows some variations of electrolytic capacitors.

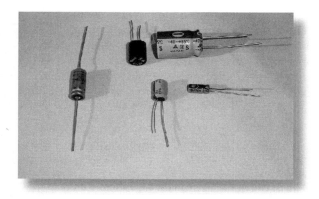

F𝗜ɢᴜʀᴇ **3.17** Electrolytic capacitors.

Electrolytic capacitors are always marked in some way — usually with a + symbol to indicate the positive component lead. Sometimes however, positive is marked with a ridged or ringed end on the component body.

Axial vs radial
Another way in which capacitors are categorized has to do with their shape, and is shown in Figure 3.17. Capacitors which are long, and have their component leads exiting from opposite ends of the body are known as *axial-leaded* capacitors. Capacitors where both component leads exit from one end of the capacitor body are known as *radial-leaded* capacitors.

Non-electrolytic capacitors
There are several categories of non-electrolytic capacitor, illustrated in Figure 3.18. Most of them are known simply by the substance that is used to make up their dielectrics. Common capacitors are *polycarbonate, polyester, polystyrene, ceramic, mica,* and *polypropylene.*

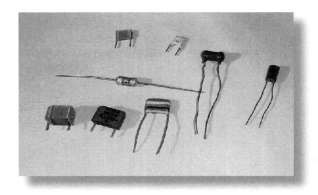

FIGURE 3.18   A variety of non-electrolytic capacitors.

Readers who have been paying attention might be wondering why it's necessary to use electrolytic capacitors at all if they're *that* awkward to use. Well, in theory, it doesn't matter which category of capacitor (electrolytic, non-electrolytic, or which type of dielectric) is used in any given circuit — a 47 nF polycarbonate capacitor has exactly the same capacitance as a 47 nF ceramic capacitor so, indeed, may technically replace it. However, the material used as a capacitor's dielectric has an effect on the capacitor's body size, shape, and component lead-out, so a capacitor of one dielectric simply may not fit into the space in a printed circuit board allocated to a capacitor of another dielectric. Also, in all but a few cases, it would simply not be possible to produce a large value capacitor from a non-electrolytic dielectric, without the resultant capacitor being huge.

Another thing to bear in mind is that the dielectric used (of both electrolytic and non-electrolytic capacitors) can have a significant effect on the component price. For example, *mica* capacitors (sometimes also known as *silvered mica*) are much more expensive to produce than, say, polycarbonate capacitors.

Finally, the dielectric has an effect on the maximum voltage a capacitor may be used at. Capacitors are rated (along with their actual capacitance) with a *working voltage* — which, fairly obviously I hope, should not be exceeded. It is up to the circuit designer to specify what the actual working voltage of the capacitor is. It is up to the constructor of the circuit to ensure that capacitors of at least this working voltage are used.

Variable capacitors

Like resistors and their variable counterparts (potentiometers), capacitors are available in variable forms. Usually though, variable capacitors are in preset form, set up when the electronic device is first made, then rarely touched afterwards. Figure 3.19 shows some examples.

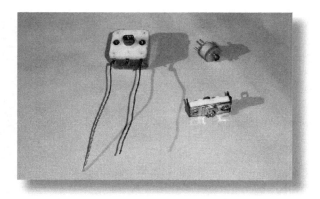

FɪɢᴜʀᴇE 3.19   Variable capacitors.

Capacitor symbols

The symbols used for capacitors in circuits are shown in Figure 3.20, where Figure 3.20(a) shows the symbol for an electrolytic capacitor (note the positive lead is marked +), Figure 3.20(b) is the symbol of a non-electrolytic capacitor, and Figure 3.20(c) is that of a variable capacitor. They are fairly self-explanatory, graphically portraying the two plates of a capacitor separated by a dielectric.

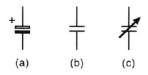

(a)   (b)   (c)

Fɪɢᴜʀᴇ 3.20   Symbols for capacitors: (a) electrolytic; (b) non-electrolytic; (c) variable.

Capacitor colour code

Very often, capacitors are marked with their value on the capacitor body. However, some capacitors are marked with a colour code in the same way that resistors are marked (see Figure 3.21), and the code (detailed in Tables 3.3 and 3.4) follows the same general method except that the value is marked in picofarads, not ohms.

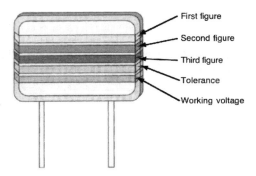

FIGURE 3.21 A capacitor marked with a colour code, showing the bands and their meanings according to the colour code of Tables 3.3 and 3.4.

Table 3.3 The capacitor colour code.

Colour	Band 1 1st figure	Band 2 2nd figure	Band 3 3rd figure	Band 4 tolerance	Band 5 working voltage (see below)
Black	0	0	x1	20%	
Brown	1	1	x10	1%	
Red	2	2	x100	2%	
Orange	3	3	x1000	—	
Yellow	4	4	x10,000	—	
Green	5	5	x100,000	5%	
Blue	6	6	x1,000,000	—	
Violet	7	7	—	—	
Grey	8	8	—	—	
White	9	9	—	10%	

Remember that to convert from picofarads to nanofarads simply divide by 1000. To convert from picofarads to microfarads divide by 1,000,000.

Band 5 of a colour coded capacitor gives its working voltage (ie, the highest voltage that should be applied to it). This does, however, vary with the type of capacitor. Table 3.4 lists capacitor types affected and gives the voltages according to the band colour.

Table 3.4   Capacitor types and working voltages according to the fifth band of the colour code of Table 3.3.

Capacitor type	White	Yellow	Black	Green	Blue	Grey	Pink	Red
Tantalum	3	6.3	10	16	20	25	35	
Polycarbonate		400			630			250
Polyester		400						250

Capacitor number/letter codes

Apart from colour coding, capacitors can also be marked with printed letters and numbers. Often these are fairly self-explanatory — a capacitor marked with the term 8p2 is obviously going to be an 8.2 pF device. However, there are some not-so-obvious variations. For example, a 680 pF capacitor is sometimes marked n68 — which means that it is 0.68 nF (which of course is the same as 680 pF). Sometimes a capacitor may be marked with just three numbers — say, 333. In such a case the first two digits indicate the first and second digits of the value, while the third digit indicates the multiplier (ie, number of zeroes to add at the end). So, the example of the capacitor printed with the number 333 would have a value of 33,000 pF, which is 33 nF.

Tolerances of capacitors are usually not very tight. Often 20% is an acceptable figure, although some capacitors with tolerances of 5% are available. Occasionally, tolerances are marked on a capacitor's body with a letter, such as J (which means 20%) or K (10%) somewhere alongside the value in numbers.

Semiconductors

There are many types of component which are grouped together and known as semiconductors. We'll take a look at some of them now, but it's important to remember that the category is huge, and there are many more semiconductors than we can possibly hope to cover here. However, there are some that are just *so* important that we can not afford to ignore them.

Diodes

Diodes are the first important component in the category known as semiconductors. Indeed they are the simplest semiconductor device. They get their name from the basic fact that they have two electrodes (di — ode, geddit?). One of these electrodes is known as the *anode*: the other is the *cathode*. Figure 3.22 shows the symbol for a diode, where the anode and cathode are marked. Figure 3.23 shows some typical diode body shapes — in which cathodes are identified as the banded or shaped ends.

The name *rectifier* is often used in place of *diode*. Strictly speaking, a rectifier is a diode that is used in high power applications, say, a mains power supply that converts mains alternating current to a low voltage direct current — the conversion from alternating current to direct current is called *rectification*. However, the two names are otherwise synonymous.

Anode   Cathode

FIGURE 3.22   The circuit symbol for an ordinary diode.

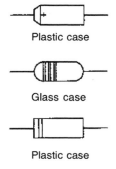

Plastic case

Glass case

Plastic case

FIGURE 3.23   Some typical diode body shapes.

Figure 3.24(a) shows a diode whose anode is positive with respect to its cathode. Although we've shown the anode as positive with a + symbol, and the cathode as negative with a – symbol, they don't necessarily have to be positive and negative. The cathode could for example be at a voltage of +1000V if the anode was at a greater positive voltage of, say +1001V. All that needs to occur is that the anode is positive with respect to the cathode.

Under such a condition, the diode is said to be *forward biased* and current will flow, from anode to cathode.

When a diode is *reverse biased* ie, its cathode is positive with respect to the anode, no current flows, as shown in Figure 3.24(b). Obviously, something happens within the diode which we can't see, depending on the polarity of the applied voltage, to define whether current can flow or not. Just exactly what this something is, isn't necessary to understand here.

We needn't know any more about it here because we're only concerned with the practical aspects at the moment; and all we need to remember is that a forward biased diode conducts, allowing current to flow, while a reverse biased diode doesn't. Which way you should insert a diode into a particular circuit depends totally on the diode and the circuit. All you have to do is insert it the specified way round.

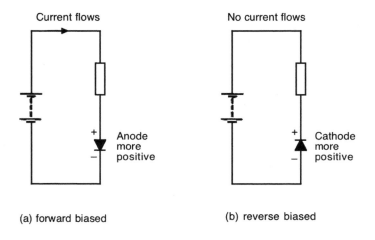

(a) forward biased

(b) reverse biased

Fᴵɢᴜʀᴇ 3.24   **Simple circuits for: (a) forward biased and (b) reverse biased diodes.**

Unlike resistors and capacitors which have particular values, diodes vary only in parameters like the maximum current and maximum voltage they can be used with, and as such are usually specified by a type number — something like 1N4001 or 1N4148. Generally, diodes are marked with their type number, which makes it relatively easy to see what type a diode is. There is one important exception though, in that some diodes are colour banded in a similar way to resistors and capacitors, and use the same colours to indicate digits. However, the four colour bands merely indicate the four digits in the diode's type. Probably the only diode you'll see type-marked in this way will be the very commonly used diode — the 1N4148, which has bands coloured *yellow brown yellow grey* to indicate 4148.

Bridge rectifier
One use of diodes is as rectifiers — devices that convert alternating current to direct current — in power supplies. Usually (but not always), four diodes are connected together to create what is known as a *bridge rectifier*. These can be constructed from individual or discrete diodes, or obtained as a complete unit which mounts directly into the printed circuit board. Several varieties of these complete bridge rectifier exist, and a few are shown in Figure 3.25. Care must be taken when inserting a bridge rectifier into a printed circuit board that it is the correct way round. Markings on the device will indicate the two alternating current input connections (usually with the symbol ~), and positive (+) and negative (–) output connections. Like discrete diodes, the two important considerations when using bridge rectifiers are the maximum working voltage (actually the maximum alternating voltage applied across it) and the current required from it.

FIGURE 3.25   Various bridge rectifiers.

Transistors
The next important step up the semiconductor evolutionary chain is the *transistor*. Invented over 50 years or so ago, it has been the single most important electronic device ever to have been made. All subsequent electronic devices have been developed from the transistor, and even

the most complex of devices such as a computer relies totally upon it still. Nowadays, however, individual transistors are rarely used. Instead several (hundreds, thousands, even millions!) are often combined into complete devices known as *integrated circuits* (we'll look at integrated circuits shortly) and used almost without knowledge or care of what is inside them.

Despite being used rarely, it's still just possible that you'll come across an individual transistor, so we need to look at them now, and fully understand them, so that you can also see how they are used in more complicated devices.

A transistor is shown in Figure 3.26. From the photograph, you'll see that it's pretty small and has three terminals. These terminals are called *base*, *collector* and *emitter* (often shortened to b, c and e). When you use transistors in electronic circuits it is essential that these three terminals go the correct way round.

FIGURE 3.26   A transistor next to a UK one penny coin.

The problem arises in that different transistors have different body shapes (see Figure 3.27) and different terminal leadouts, so to make sure that the transistor is the correct way round, manufacturers supply them in a standard range of body shapes, and each body shape has a standard terminal lead-out. So, by knowing what body shape a transistor is you can work out its terminals from a chart. Common varieties you'll even learn by heart automatically, just by using them a couple of times.

FIGURE 3.27   A selection of transistors, showing some of the variance in body shapes available.

Figure 3.28 illustrates leadout details for a selection of common transistors. From those shown, you can see that it's easy to work out terminals. For most transistors, leadouts are conventionally given as a plan view looking at the transistor's underside ie, with the transistor's leads pointing towards you. The exceptions are certain types of power transistors, where you view the transistor looking towards the metal pad of the transistor as an elevation view.

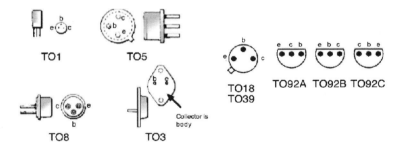

FIGURE 3.28   Leadouts of some commonly used transistors. Manufacturers of transistors can provide details of all their devices.

There are several different types of transistor, the most popular being what is called the *bipolar* transistor. These are constructed in two different ways, as three layers of semiconductor material — as a thin layer sandwiched between two thicker layers, as shown in Figure 3.29. The semiconductor layers are of two types — called *N-type*, and *P-type*. So the transistor can be made with a thin P-type layer sandwiched between two N-type layers, or with a thin N-type layer sandwiched between two P-type layers — which way defines the transistor's type as either an NPN transistor, or a PNP transistor.

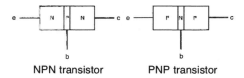

NPN transistor   PNP transistor

Figure 3.29   **An NPN and a PNP transistor make-up.**

The circuit symbols for the two types of bipolar transistor — NPN and PNP — are shown in Figure 3.30, along with their corresponding terminals. Usually, in circuit diagrams, terminals are not given, and it's taken as read that you will know which terminal is which.

Although the internal working of transistors is pretty complex, we can look on them quite simply — they are electronic switches, in which a small current through the base terminal creates a much larger current through the collector and emitter terminals. The small current through the base terminal is by convention called the *base current*. The much larger current through the collector and emitter terminals is called the *collector current*.

NPN transistor   PNP transistor

Figure 3.30   **Symbols of NPN and PNP transistors.**

The way bipolar transistors work is illustrated in Figure 3.31(a) for an NPN transistor, where the base current flows into the transistor's base causing a large collector current into the collector. Figure 3.31(b) shows a small base current out of a PNP transistor's base causing a large collector current out of the collector terminal.

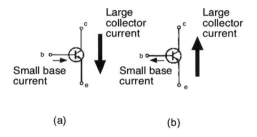

FIGURE 3.31 Showing how both NPN and PNP bipolar transistors work — a small current causes a large collector current.

Field effect transistors

Another major group of transistors are called *field effect transistors* (abbreviated to FET), of which there are two main types: the *junction gate field effect transistor* (JUGFET, or J-FET) and the *metal oxide semiconductor field effect transistor* (or MOSFET).

Like bipolar transistors, FETs can be of N- or P-types, but are actually known as N-channel or P-channel types. The three terminal connection leads are the *gate*, *drain* and *source*, which are roughly the equivalent of base, collector and emitter. MOSFETS are available with two gate terminals, and are called *dual-gate MOSFETS* (or DG MOSFETS).

Figure 3.32 shows the circuit symbols of a range of FETs.

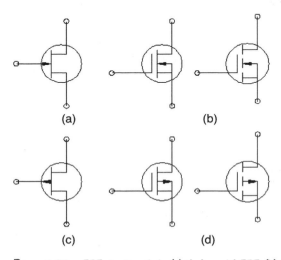

FIGURE 3.32 FET circuit symbols: (a) N-channel J-FET; (b) N-channel MOSFETs; (c) P-channel J-FET; (d) P-channel MOSFETs.

Thyristors

It's sometimes necessary to control current flow in circuits. A specific device to do this is the *silicon controlled rectifier* (or SCR), which also goes by the name *thyristor*. Much like a transistor in appearance, with the terminal *anode* (A), *cathode* (K) and *gate* (G), the thyristor operates like a rectifier or diode that can be switched on or off. Figure 3.33 shows the circuit symbol of a thyristor, while Figure 3.34 shows some common thyristor leadouts.

FIGURE 3.33   Thyristor circuit symbol.

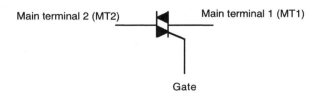

FIGURE 3.34   Some thyristor leadouts.

Triacs

Where thyristors are used to control direct current, *triacs* are used to control alternating current. Their terminals are known as *main terminal 1* (MT1), *main terminal 2* (MT2) and *gate* (G). Circuit symbol for a triac is shown in Figure 3.35, while Figure 3.36 shows some common triac leadouts.

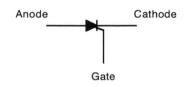

FIGURE 3.35   Triac circuit symbol.

FIGURE 3.36   Some triac leadouts.

Integrated circuits

There are many, many different types of *integrated circuit* (abbreviated to IC) — far too many to describe in this book. Fortunately, at least for beginners to the subject of electronics, it's possible to generalize somewhat and identify perhaps a handful of integrated circuit types that are worthy of note.

For a start, most integrated circuits you'll come across are in a particular body shape — the *dual-in-line* integrated circuit (or DIL IC). These have two rows of terminal pins which are used to connect the internal circuit of the integrated circuit into an electronic circuit.

An example of a very common integrated circuit (the 555, as it happens) is shown in Figure 3.37, where you can see that it's got a total of eight terminal pins, ordered in two four of four. The rows of terminal pins of standard dual-in-line integrated circuits like the 555 are spaced 0.3 inch apart, and pins in the rows are 0.1 inch apart.

FIGURE 3.37   A 555 integrated circuit.

All dual-in-line integrated circuits like the 555 have their terminal pins numbered in the same way — pin 1 is top left, and all other pins follow in an anti-clockwise direction. All that's needed is to know which pin is number 1. Unfortunately, here, there is some difference between integrated circuits manufacturers. Figure 3.38 shows the most common methods of manufacturer markings for dual-in-line integrated circuits — note that (unlike transistors and their associated semiconductor components) integrated circuit pin diagrams are as viewed from the top of the integrated circuits.

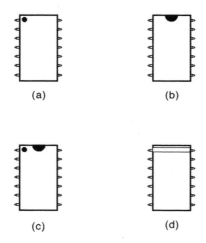

(a)　　　　　　　　(b)

(c)　　　　　　　　(d)

FɪɢURE 3.38　Terminal pin 1 on dual-in-line integrated circuits is indicated by: (a) a dot; (b) a notch; (c) a dot and notch; (d) a line. Manufacturers vary, so you need to know the various methods.

Integrated circuit types

Integrated circuits contain highly miniaturized electronic circuits — they can, in fact, contain literally thousands of tiny transistors inside their bodies. It should come as no great surprise then to learn that there are literally thousands of different types of integrated circuits, each one designed to do a slightly different job to its neighbours. The exact job an integrated circuit does depends almost totally on the transistors that make up the circuit inside it. As such, it might seem difficult to find out what all these many thousands of types of integrated circuits do.

Fortunately though, integrated circuits fall very roughly into two highly generalized main categories — *analog* (sometimes known as *linear*) and *digital*. Analog integrated circuits are those that can be used in the likes of audio circuits, or some forms of control circuits for example. Digital integrated circuits are those that would be used in the likes of computing circuits, and also some forms of control circuits for example.

As you might expect, there is some overlap between the two categories, which blurs the distinction between analog and digital integrated circuits somewhat.

Typical analog integrated circuits

Perhaps the most common forms of analog integrated circuits are those that comprise the family of integrated circuits known as *operational amplifiers* or *op-amps*.

Operational amplifiers are often illustrated in an electronic circuit diagram with the symbol shown in Figure 3.39(a), where it can be seen that the device has two inputs (marked + and –), and an output.

It's also common to see an operational amplifier illustrated in an electronic circuit diagram inside its integrated circuit body, showing its terminal pins as part of the circuit, as shown in Figure 3.39(b).

Sometimes, the operational amplifier would be shown in a circuit diagram with its actual terminal pins numbered, but the integrated circuit body is not actually shown (Figure 3.39c). All three methods are strictly correct, and it really just depends on the level of detail required in a circuit diagram, along with personal preference, as to which method is used.

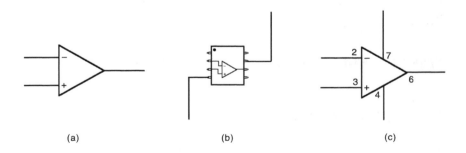

(a)   (b)   (c)

FIGURE 3.39   Circuit symbols for integrated circuits: (a) just the op-amp, with no pin numbers; (b) showing the op-amp complete with integrated circuit 'body' and pin numbers; (c) the operational amplifier with pin numbers.

▷ **FYI** ◁

While the symbol for an op-amp is relatively simple, you should realize that the op-amp itself may contain many hundreds of components inside the integrated circuit. Treating them collectively as a single component called an op-amp simply makes it easier for us to get our heads round it. An op-amp is a sort of 'black box' electronic component — we don't need to know how it's made up, just how it works. There are many other black box electronic components.

Many integrated circuit op-amps have identical terminal pins, so it's possible that one op-amp can be replaced by another in the same electronic circuit. But there are some differences, so before you use an op-amp, you should refer to manufacturers' data sheets for the op-amp to make sure it will fit in, and work in, the electronic circuit you are building. That being said, most integrated circuit manufacturers make a range of general-purpose op-amps that are sufficiently close in relevant characteristics that they may be classed as identical. For instance, most integrated circuit manufacturers make an op-amp called a 741. A 741 from one manufacturer should work in a 741 designed for another manufacturer's 741. For reference, Figure 3.40 (left) shows a 741 op-amp — it's in a standard 8-pin, dual-in-line integrated circuit format, while Figure 3.40 (right) shows the pin-out diagram for a 741.

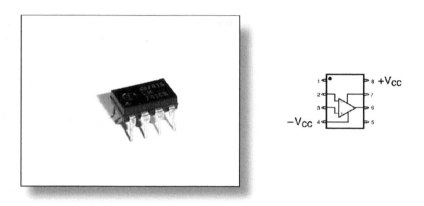

Fɪɢᴜʀᴇ **3.40**　A 741 op-amp: (left) photograph; (right) pin-out diagram.

Typical digital integrated circuits

Perhaps the simplest forms of digital integrated circuits are those that make up digital gates. Like analog op-amps, these usually have a simple black box symbol, but in reality are made up from many individual transistors inside the integrated circuit. Common logic gates are shown in Figure 3.41. There are, however, many more types of logic circuits available in integrated circuit form. Also, like analog op-amps, logic gates are commonly shown in an electronic circuit with their pin numbers shown.

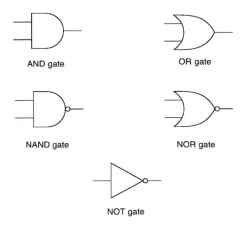

FIGURE 3.41   Common logic gate symbols.

There are several families of digital integrated circuits, generally based on their construction techniques. How the actual gates or logic circuits inside each integrated circuit in each family are connected through their respective terminal pins is defined in standard digital integrated circuit pin-out diagrams. Some digital integrated circuit pin-out diagrams are shown in Figure 3.42 (for digital integrated circuits in the 7400-series), and in Figure 3.43 (for digital integrated circuits in the 4000-series). These are the most common digital integrated circuit families.

Batteries

Electronic circuits need power of some sort to function. Power is sourced from either a power supply (usually mains-powered) or — particularly for portable equipment designs — batteries. Occasionally, a single cell might be sufficient to power an electronic circuit — the symbol is shown in Figure 3.44 (a) — but more usually multiple cells, or power pack batteries are used — the symbol of both is shown in Figure 3.44 (b). Note that the voltage should be shown beside either symbol.

Where cells or batteries are used to power electronic circuits, some form of connector must be used to clip the cell or battery into place. Various battery connectors are shown in Figure 3.45. The type used depends on the cells or batteries required, the circuit itself, and also the housing used to hold all the electronic circuit parts (including the cells or battery).

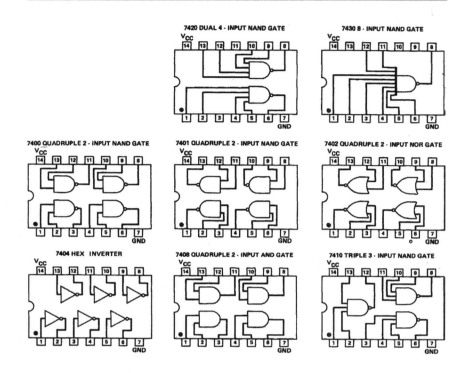

FIGURE **3.42   Some pin-out diagrams for members of the 7400-series of digital integrated circuits.**

Indicators

Often, it's useful or necessary to show when an electronic circuit is turned on. This is usually done with an indicator of some sort — either a bulb, a neon, or most often a LED.

Bulbs

Not used very often, low-voltage bulbs can give an effective indication of an electronic circuit's function. However, problems associated with mounting the bulb normally preclude their use. Some bulbs and mounting methods are shown in Figure 3.46.

Neons

Largely for the same reasons we don't often use bulbs, neons aren't often used either — with the possible exception of some mains switches (see later) that contain an integral neon bulb to show when an electronic circuit's power supply is turned on. Figure 3.47 shows an example of such a mains switch.

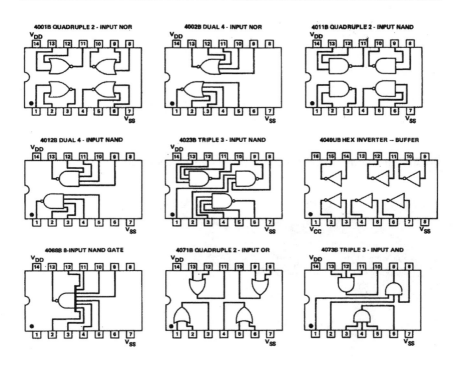

FIGURE 3.43   Some pin-out diagrams for members of the 4000-series of digital integrated circuits.

LEDs

Light emitting diodes (LEDs) nowadays form the usual and best method of indication. They are easily used, and easily and cheaply mounted onto housings' front panels, so have become the preferred method. They are — as their name suggests — a particular form of diodes, and must be used with the same precautions as other diodes (see the section on diodes, earlier in this chapter), but when they are forward biased in a circuit emit light. Normally one side of the LED is marked in some way to indicate cathode. Different colours of LEDs are available (green, red, blue, white), and some forms emit light of different colours depending on connections. Figure 3.48 shows a typical round LED, although rectangular LEDs are also available.

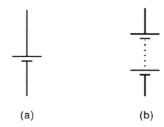

FIGURE 3.44   Battery symbols (a) single cell (b) multiple cell or powerpack.

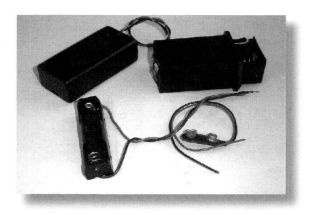

FIGURE 3.45   A selection of cell and battery connectors.

FIGURE 3.46   Bulbs and panel-mounted bulb holders.

FIGURE 3.47   A mains switch with an integral neon bulb (this particular switch is used in the Bench Power Supply project in Chapter 8).

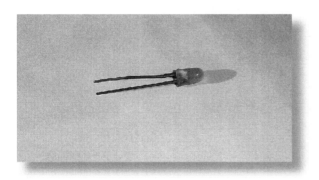

FIGURE 3.48   A typical LED.

Switches

There are many forms of switches used in electronic circuits, and they take many different shapes and sizes. They are all similar however, in that they are all mechanical devices which are physically operated by the user. Their main differences lie in how the internal contacts are arranged, and what form of mechanical device is operated by the user to switch contacts.

Contacts

The simplest switch has two contacts, and a simple on/off mechanism. Its circuit symbol is shown in Figure 3.49 (a). As it can be used in only one connection within a circuit, it is called *single-pole*. Also, as it can be only on or off, it is called *single-throw*. So, in full, it is known as a *single-pole, single-throw* switch (shortened to *SPST*).

Where the switch has a third contact, so the central contact can be switched to either one state or another, it is known as a *single-pole, double-throw (SPDT)* switch, and is occasionally called a *changeover switch*, and its circuit symbol is shown in Figure 3.49 (b).

Commonly, switches can be joined (or *ganged*) together in one housing, to allow more than one connection to be switched at the same time. The simplest of these switches would be a *double-pole, single-throw (DPST)* switch, the circuit symbol of which is shown in Figure 3.49 (c). The broken line between the two individual mechanisms of the circuit symbol indicates that both mechanisms are mechanically joined, and as one switch operates the other does too.

Circuit symbol of a *double-pole, double-throw (DPDT)* switch is shown in Figure 3.49 (d), where the two joined mechanisms are of double-throw format. Again the broken line joining the individual mechanisms indicates that both mechanisms operate in unison.

While these four switch types shown in Figure 3.49 (a) to (d) are obviously not the only switch types, they form by far the most common types you will come across in electronics.

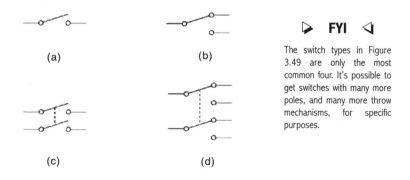

(a)   (b)

▷ **FYI** ◁

The switch types in Figure 3.49 are only the most common four. It's possible to get switches with many more poles, and many more throw mechanisms, for specific purposes.

(c)   (d)

FIGURE 3.49   Four main switch types: (a) single-pole, single-throw (SPST); (b) single-pole, double-throw (SPDT); (c) double-pole, single-throw (DPST); (d) double-pole-double-throw (DPDT).

> **Take Note**
>
> The contacts of switches of all types have only limited voltage and current ratings, and you must make sure that you do not exceed these contact ratings when using switches. If the circuit a switch is used in is battery-powered, it is unlikely that contact ratings will be exceeded, but if the switch is to be used to switch mains power then contact ratings should be carefully checked before use.

Mechanics

The most common mechanics used in switches in electronics are: slide, toggle, push, and rotary. Examples of all are shown in Figure 3.50. Each type may have a number of deviants, and is usually available in a range of sizes too. Switches may also be incorporated with other functions, such as a potentiometer (see page 31), or indicators (see page 58). Also, each mechanism type can have minor variants in operation — for example push switches can be *push-to-break* or *push-to-make*, and *switched* action or *momentary* action — so it's not hard to see that there are many hundreds of switch types to choose from.

FIGURE 3.50   A range of switches, according to mechanism types, from left to right: slide, toggle, push, rotary.

Relays

Relays are just switches that are operated electro-mechanically. Thus, where an ordinary switch may be operated by you pushing a button, or flicking a toggle, a relay is operated by merely applying an electric current to an electro-magnetic coil. This current creates a magnetic field that flicks the relay's internal switch mechanism. The advantage is pretty obvious — an electronic circuit can operate the relay automatically, without human intervention. A circuit symbol for a single-pole, double-throw relay is shown in Figure 3.51.

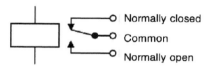

Normally closed
Common
Normally open

FIGURE 3.51   Circuit symbol of a single-pole, double-throw relay.

Generally, relays are mounted on an electronic circuit's printed circuit board, so are quite small. Several variants exist, and usually it is simply a matter of choosing the relay to suit in terms of number of switch poles and throws, contact ratings, coil operating voltage and current, and size. An example of a relay — a miniature, printed circuit board mounting relay — is shown in Figure 3.52.

FIGURE 3.52   A miniature, printed circuit board mounting relay.

Loudspeakers and other sound output devices

In an audio circuit, it's often required that we can hear the audio signal. The usual way to do this is either with an in-built loudspeaker, or using earpieces.

Loudspeakers built in to electronic circuits are usually fastened to the housing of the circuit, and once in place are pretty robust. However, they do require some precautions in handling. They work because a cone of some material is vibrated electro-magnetically (due to an electric current passing through an attached coil of wire that is situated in a magnetic field). Problems can arise because the cone is easily damaged or split if mishandled. A couple of examples of loudspeakers for use with electronic circuits are shown in Figure 3.53.

FiGURE 3.53   Loudspeakers typically used in electronic circuits.

Sometimes, while audio sound is not outputted from a circuit, some form of noise is — say, a buzz, or beep. There are several forms of devices for doing this, but only two main groups: *buzzers* and *sounders*.

Buzzers are generally electro-mechanical or piezo-electric sound sources which, when a voltage (generally of direct current) is applied across their terminals, vibrate at a fixed frequency, causing a sound. They come in various shapes and sizes and amplitude levels, some of which are shown in Figure 3.54. The circuit feeding the buzzer merely needs to switch a current to the buzzer to make it work.

Sounders are more like loudspeakers in operation (some are shown in Figure 3.54), and require some form of alternating current signal. Unlike loudspeakers, however (which can successfully operate over a range of frequencies — so can reproduce audio quite well), sounders require a fairly specific frequency signal to generate the loudest sound possible. The circuit feeding the sounder must produce an oscillation at the frequency required by the sounder, to make it work.

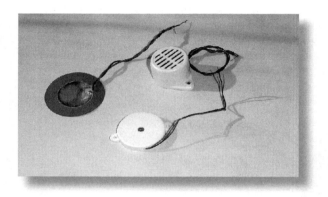

FIGURE 3.54   A buzzer with two types of sounders.

Microphones

Microphones are, in essence, the opposite of loudspeakers and sounders. They convert a sound into an electrical signal, for use by an electronic circuit. They are common in two main forms: *dynamic* and *electret*. Dynamic microphones are just like loudspeakers in reverse; they feature a diaphragm (that is much like a loudspeaker's cone) which vibrates as sound waves hit it, connected to a coil of wire that moves through a magnetic field, producing electric current.

Electret microphones (sometimes known as *condenser* microphones) operate in a vaguely similar way in that they have a polymer diaphragm that creates an electric signal as it vibrates due to impinging sound. However, this signal is too small to be used raw, so they also include an amplifier inside the microphone housing to amplify the signal. While this produces an extremely high quality and small microphone, the downside is that the amplifier must be powered somehow, but a simple low-voltage connection is all that is required (and, as most electronic circuits are powered by a low voltage, extremely easy to include).

The circuit symbol for a microphone is shown in Figure 3.55, while Figure 3.56 shows microphones of the two main types.

FIGURE 3.55   Circuit symbol for a microphone.

FIGURE 3.56   An electret microphone, and a dynamic microphone.

Coils and inductors

Technically, any coil of wire that has a current applied to it becomes an inductor. So, any electronic component that comprises a coil can also be classed as an inductor. That being said, it's not often that simple coils are used in an electronic circuit (normally only in radio frequency circuits). In most instances, you may even not realize that the component you use features an internal coil.

Examples of coils in common use that you may use are transformers (used, say, in mains power supplies), loudspeakers, and relays (see earlier).

Connectors

Any electronic circuit depends on connectors of many types: to get signals into and out of the circuit; to connect to controls that are on a housing's front panel; and to obtain power from a battery or power supply. The range of connector types is huge, and can't fully be covered here, but we can at least look at a range of common connectors you will almost certainly come across. Connectors are generally (as far as we're concerned here) used either on a printed circuit board, or on the housing used to house a printed circuit board.

Some typical plug and socket connectors for use on printed circuit boards are shown in Figure 3.57. Some typical connectors for mounting on a circuit's housing are shown in Figure 3.58.

FIGURE 3.57   Typical printed circuit board connectors.

FIGURE 3.58   Connectors used on housings for front- or rear-panel mounting.

Integrated circuit holders

Where integrated circuits are to be fitted to printed circuit boards, it's common not to solder them in. Integrated circuits are, of course, complex, miniaturized devices. Excess heat when soldering may damage them (although some are less sensitive to heat damage than others), so an alternative is to use *integrated circuit holders* (sometimes called *integrated circuit sockets*), which *are* soldered into the printed circuit board. Integrated circuits are then push-fitted into the holders afterwards.

Figure 3.59 (left) shows an integrated circuit holder in place on a printed circuit board, while Figure 3.59 (right) shows the holder with an integrated circuit fitted. Integrated circuit holders are sized the same way that integrated circuits themselves are, so you would use an 8-pin holder for an 8-pin integrated circuit, a 14-pin holder for a 14-pin integrated circuit, and so on. Holders are not expensive — about 15p each — so it's worth buying a few of each common size (8-pin, 14-pin, 16-pin) for your component box.

FIGURE 3.59   Using integrated circuit holders: (left) the holder itself, in position in a printed circuit board; (right) showing an integrated circuit in the holder.

Terminal pins

One of the main problems with a printed circuit is how to connect any external controls, switches or connection sockets. Often, it's possible and easy to use a connection lead, from the component side of the board to the track side, soldered into place at a junction in the copper foil track. Indeed, Chapter 5 shows in detail how to do this. However, if a printed circuit board has been mounted in a housing (see Chapter 6) then it's not easy to get to both sides of the board to do this.

It's therefore a good idea to use terminal pins. These push through from the copper track side of the printed circuit board to the component side. They are then soldered to the copper track. Thus, when the printed circuit board is mounted in its housing, it's still easy to solder a connecting lead to the pin connection (which is on the component side ie, the upper side, of the board). Chapter 5 shows how to solder a connecting lead to a terminal pin, but in order to do that, of course, you need to buy some terminal pins. They are usually sold in packs of 100 and cost only a few pounds a pack, so are well worth the investment. Various sizes are available, and various types, but the most common 1 mm diameter, single-sided, pre-tinned type is all you'll need, unless you are constructing a circuit that calls for anything more unusual.

> **Take Note**
>
> Don't buy gold-plated terminal pins — they are much more expensive, and in all but the most critical of situations, provide no benefit.

Figure 3.60 shows terminal pins, soldered in place in a printed circuit board.

FIGURE 3.60 Terminal pins in a printed circuit board — the pins are push-fitted into the board, then soldered in place.

▷ **FYI** ◁

Terminal pins have a knurled ridge that holds the pin in place in the printed circuit board. The knurled ridge makes them a little hard to push into place however. Pliers, or a hammer, will do the job if you are careful. However, you may think it worth buying a terminal pin pushing tool to help with this job — see Chapter 2.

Printed circuit board links

While hardly being an individual component, it's best to take a look at the use of printed circuit board links. These are used on a printed circuit board if a connection in the circuit needs to be made, but the design of the copper foil track is such that it cannot be made using the copper foil.

In such cases, it's common to use a small link of wire that jumps from one connection point to another on the printed circuit board.

Many connections in the copper foil track can be run underneath the link, so it forms a useful method of bridging parts of the printed circuit board.

Figure 3.61 shows a printed circuit board with several links in place — the board shown is actually the Heart Flasher project printed circuit board, detailed in Chapter 8.

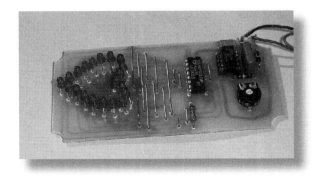

FIGURE 3.61   Links on a printed circuit board.

Printed circuit board links *can* be made with insulated connection lead, but this tends to make a rather messy board. Instead, it's best to use uninsulated single-strand wire. Being rigid, links don't bend and become untidy the way insulated wire would, and there is little risk of links touching other links or neighbouring components. Buy a reel of single-strand wire, then cut short pieces off to make links — see Chapter 5 for details.

Take Note

When designing a printed circuit board which requires links, make sure that links run parallel – in this way there is no danger they will touch other links close by.

4

Printed circuit boards

Most of the projects you'll ever want to make up from electronics magazines or electronics books like this one are constructed on printed circuit boards (PCBs). Without any doubt printed circuit boards help you make really neat and reliable projects, simply because they hold all the components of the projects in a fixed and long-standing way, with strong soldered joints giving the good electrical connections required in the circuits. You've only to take the back off any modern electronic appliance such as your television, radio, hifi amplifier, computer (carefully — and after you've unplugged it from the mains and assuming it's yours!) to see that it's not just book and magazine projects which use printed circuit boards, but manufacturers of electronic appliances do, too. Figure 4.1 shows a printed circuit board in a typical modern appliance.

FIGURE 4.1   A printed circuit board - the type you might find in a television or other modern electronic appliance.

There are loads of benefits when using printed circuit boards:

NUMBER 1 Positioning components. Each component in the circuit has its own special little place on the printed circuit board — if you've forgotten to insert it, it's fairly obvious there's something missing because there is a space and a couple of holes

NUMBER 2 The components fit close up to the board. Component leads go through holes in the board to the other side, where they are soldered to strips of copper — in other words, the components are rigidly held so they can't easily work loose, and there's not much chance of one component lead shorting against another

NUMBER 3 Size. Yet another benefit is simply the sheer compactness of the completed assembly — as connections are made with copper tracks on the underside of the board (some printed circuit boards have copper tracks on both sides — and are called, naturally enough, double-sided printed circuit boards) and as components can be very close together without the risk of short circuits, the size of the printed circuit board need be very little more than just the total surface area of the components.

If you have an old valve radio or television around the place, you can compare equipment with modern printed circuit boards to the method used then; where all the components were simply strung between connection points on 'tag strips' underneath a metal chassis, in a seemingly haphazard arrangement. While such methods served their purpose in their day, no one can doubt that printed circuit boards are infinitely better.

Copper-clad printed circuit board

So what exactly is a printed circuit board, what's it made of, how's it made, and how do we set about getting it?

Well, the photograph in Figure 4.2 shows a small piece of printed circuit board, directly as it would be obtained from a printed circuit board manufacturer, or as you might buy from an electronics components shop. This small piece would have been cut from a much larger sheet (typically a few metres square). In its raw state like this, the printed circuit board is

Figure 4.2   Raw printed circuit board, comprising a thin layer of copper on a insulating material baseboard.

called *copper-clad board* and it comprises a thin baseboard (about 1.5 mm) of insulating material such as resin-bonded paper or, more likely, fibreglass, with an even thinner layer of copper (about 0.2 mm) on one surface (or both, if it's double-sided copper-clad board). Figure 4.3 shows an enlarged cross-section of a typical piece of copper-clad board.

Copper

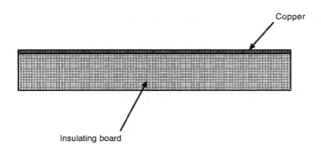

Insulating board

Figure 4.3   Enlarged cross-section of copper-clad printed circuit board

▷   **FYI**   ◁

Apart from basic copper-clad printed circuit boards like this, industrial manufacturing often makes uses of double-sided copper-clad printed circuit boards (in which a circuit is made up from copper connections on both sides of the board) and multi-layered printed circuit boards (which comprise several internal layers of copper connections as well as those on one or both surface sides of the board) — typically these would be used in only the most complex of circuits, where space is at a premium!

This copper-clad board must now be turned into a usable printed circuit board, so some way must be found to remove excess areas of copper, leaving behind only those areas which we need to connect the components to. The areas left behind form what's called the copper track or the copper foil. Figure 4.4 shows such a printed circuit board where the copper track or foil is clearly visible.

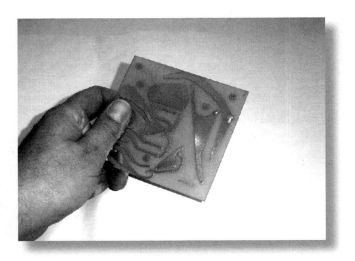

FIGURE 4.4   Once the excess copper is removed from copper-clad board, the copper track or foil remains.

Once a printed circuit board has been created, complete with a copper track, it is then a relatively simple task to drill the holes in the printed circuit board for the component leads to fit through, and then solder them in.

But we shouldn't go too far, too fast. We should really stop and think for a minute about what's just been said. Somehow a piece of copper-clad board (which consists of a thin layer of copper stuck onto the surface of a thin insulating board) must have areas of copper removed, leaving behind a copper foil which follows the foil pattern required on the printed circuit board for the project. Just exactly how do we remove the copper?

Don't worry, the answer's not quite as difficult as you might imagine — we simply dissolve it away using an acid-type etchant.

But doesn't the etchant dissolve away the copper foil, too? Well, no — because first we have to coat the areas of copper that are to make up the copper foil with something which will resist the etchant, preventing the copper underneath from being dissolved. Figure 4.5 shows the idea.

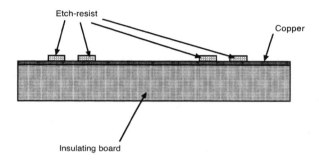

FIGURE **4.5** Coating the copper surface with an etch-resistant material means that the copper under the material is protected from attack from etchant.

After etching (Figure 4.6), the unwanted copper has been dissolved, but the copper foil under the etch-resist remains.

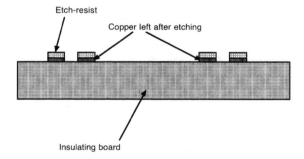

FIGURE **4.6** Once the board of Figure 4.5 has been etched, only the copper underneath the etch-resistant material remains.

After removal of the etch-resist, the printed circuit board is ready for drilling (Figure 4.7), component insertion (Figure 4.8), soldering (Figure 4.9), and finally component lead trimming (Figure 4.10).

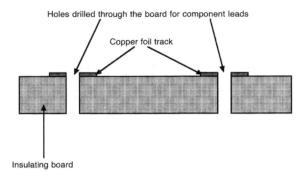

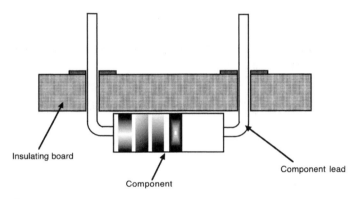

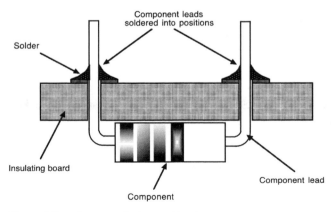

FIGURE 4.7   Holes are drilled through the board at relevant points, after the etch-resistant material has been removed.

FIGURE 4.8   Component leads are inserted through the drilled holes.

FIGURE 4.9   Component leads are soldered into position.

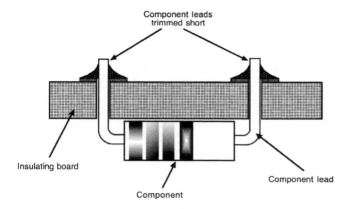

Component leads
trimmed short

Insulating board

Component lead

Component

FIGURE 4.10   Component leads are trimmed, leaving a neat soldered printed circuit board.

In summary, the steps to making a printed circuit board are:

- **OBTAIN** or design the foil pattern for the circuit

- **APPLY** etch-resist to copper clad board in the shape of the foil pattern

- **ETCH** away the unwanted copper

- **REMOVE** the etch-resist and drill the component lead holes

- **INSERT** the component leads

- **SOLDER** the component leads to the board

- **TRIM** the excess component leads.

Later in this chapter these steps are described in full, and the various methods that you can use to create your own printed circuit boards are detailed.

Other boards

There's no doubt that printed circuit boards give a much better way to
build circuits, and it's the way recommended wholeheartedly by this
book and by anyone else who knows anything about basic electronics
and who seriously wants to help people like you — the reader — to use
the best methods.

However, printed circuit boards aren't the *only* method of building
electronic circuits. So, we're going to take a short break from printed
circuit boards, to look at a couple of other methods. That way, you'll see
the differences, and see why printed circuit boards are best.

Matrix board

Made from a plain board of insulating material, with holes at 0.1 inch
horizontal and vertical distances, matrix board is merely a way of
holding components on a board. A plain piece of matrix board is shown
in Figure 4.11.

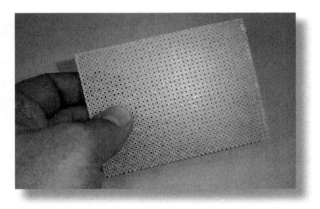

Figure 4.11   A plain piece of matrix board.

As with printed circuit boards, component leads are inserted through the
holes on matrix board, but there the similarity ends. There is no copper
on the underside of a matrix board as there is on a printed circuit board,
so all connections between components must be made by either bending
component leads to the next component or by adding extra wire to
do the same. Where component leads or wires join, the joint must be
soldered.

Overall, matrix board is useful for prototyping a simple circuit to make sure it works. It can also be useful where the circuit is being designed as it's built — it's quite easy to add extra components if a change is required, and it's also quite easy to adapt the circuit as you go. Indeed, when using matrix board *the whole layout* is usually designed as you go! However, it's not the ideal construction method for a finished electronic circuit as it is hard to replicate, and is tricky to use if the circuit is complex.

Stripboard

Stripboard is a sort of cross between matrix board and printed circuit board, so has advantages from both methods of construction. For a start, the underside (shown in Figure 4.12) has strips of copper to make connections between components. Also, the board is pre-drilled, with holes at 0.1 inch horizontal and vertical distances, just as with matrix board. As a result, it's quite easy to make a simple electronic circuit by inserting components, and soldering them in place. Also, as the copper strips are pre-defined, and unlike the earlier matrix board construction method, it's perfectly possible to design a layout on paper first, so that circuits can be replicated fairly easily. For this reason, stripboard circuit layouts are sometimes given in electronics magazines and books, particularly for simple circuits. They are not, however, ideal.

FIGURE 4.12   A plain piece of stripboard, showing the strips of copper. Pre-drilled holes in the strips mean that component insertion is straightforward, if defined.

When using stripboard it's often the case that only a couple of holes on a copper strip are used to connect components – the rest of the copper strip is then unusable. Also, where components like dual-in-line integrated circuits are to be used, the copper strips of plain stripboard would directly connect the integrated circuit's two rows of pins.

The answer is to cut the copper strip, effectively splitting it into parts. A sharp knife such as a craft knife could be used to split the copper strip, but a purpose-made tool – the spot-face cutter – is best used. To use the spot-face cutter just position it onto a hole in a copper strip, then twist it backwards and forwards two or three times, until the copper strip has been cut away (shown in Figure 4.13).

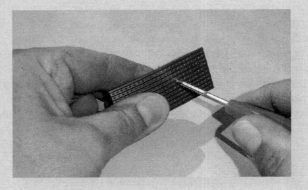

FIGURE 4.13   Using a spot-face cutter to split a copper strip on a stripboard.

There are variants of the plain stripboard, designed particularly for integrated circuits, in which the copper strips are organized to accommodate the rows of dual-in-line pins common to integrated circuits.

Nevertheless, other than being quite a useful method for the making of simple electronic circuits or for prototype circuits, stripboard is still not the ideal contructional method. Printed circuit boards remain the best method and should be the ideal you strive for.

Making your own printed circuit boards

Many people take advantage of the various printed circuit board services that electronics magazines offer, through which they can purchase ready-made printed circuit boards from the magazine publisher, complete and ready to build their projects onto. If you don't fancy making your own boards, these services are ideal.

> **Hint:**
>
> When electronics books and magazines give printed circuit boards for projects, they do so by printing what is called the foil patterns of the projects. It's the foil pattern of any particular project, of course, which is the shape of the copper to be left on the printed circuit board when the excess areas are removed. When you make a project given in a book or magazine you have the foil pattern to create the printed circuit board. If you want to design and build any other circuit, you must design the foil pattern yourself.

On the other hand, you may be trying to build a project taken from a book (like this one) or — simply — you may like to have a try making your own. Making your own printed circuit boards is certainly more rewarding than buying them ready-made — but it is a little more complex. However, it's nothing the average newcomer to electronics can't handle; as long as you follow the few simple rules which we'll now give you.

Right from the very beginning — the foil pattern

Let's assume that you have a circuit, which you may have designed yourself or somebody else may have designed it, but you have no printed circuit board foil pattern for it. Figure 4.14 is a typical circuit which you may recognize as being a multivibrator, and will do nicely for our purposes here.

At this point you need some basic tools — all of which you probably have at hand — paper (plain, tracing and graph), a pencil and a rubber, and a ruler. You'll also need a selection of one of each type of the components in the circuit eg, one resistor, one capacitor, one transistor, one integrated circuit etc, or their physical dimensions. This is because you need to know how big each component is so that when designing the foil pattern you know what distance to leave between component lead holes.

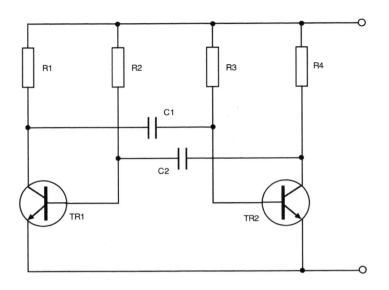

FIGURE 4.14   A simple circuit which we'll use to show how to design your own printed circuit boards.

Although this may seem daunting at first, don't worry, because if you design more than just a couple of foil patterns, you'll begin to remember the sizes of all but the most unusual components. Lay everything out on a desk or clean workbench.

Now the difficult bit (not really!): designing the foil pattern. What you have to remember here is that the components go into one side of the printed circuit board, but the copper foil making the connections is on the other side of the board.

Begin by choosing a component, fairly central to the circuit (we'll choose transistor TR1). Now draw its approximate shape (as viewed from above) onto a sheet of plain paper, as in Figure 4.15.

Just as we've done, mark where its connection leads will enter the paper 'board' and label it (TR1 in our example). Also, if it's anything other than a simple resistor or capacitor, you should mark the leads, too (e, b, and c if it's a transistor).

Now draw a few more components onto the paper, alongside the first, similarly marking them as you go (Figure 4.16). Don't be afraid to position them close together; a spacing of only 2 mm or so between components is normally ideal.

Figure 4.15   First steps: marking the first component onto plain paper. Also mark its leads – as it's a transistor, the leads should be named e, b and c to show the emitter, base and collector.

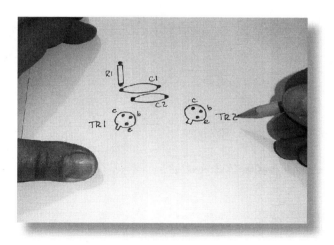

Figure 4.16   Add further components to the layout, by marking them on the paper – they can be quite close.

At this stage you may find it helpful to more or less follow the general pattern of the circuit diagram when positioning components, although this is by no means essential — it's probably just easier for a newcomer to do.

When you gain confidence in making printed circuit boards you probably won't follow the pattern of the circuit diagram at all — it really doesn't matter; as long as the connections follow the circuit diagram, the pattern can be like the Union Jack if that's what you want.

When you have drawn in a few components, begin to pencil in 'connections' corresponding to the copper track required, as in Figure 4.17, so that they will be electrically joined by these connections on the finished printed circuit board.

Don't worry if you get stuck, or even make a mistake, that's why you have the rubber! Simply rub out the error and start again.

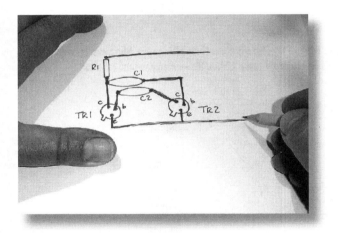

FIGURE 4.17   *Connections between component leads should be drawn in now.*

Build up the layout, putting in more and more components, and more and more 'connections'.

Remember that the finished printed circuit board will have its foil pattern on the other side to its components, so tracks can actually pass underneath components, as in Figure 4.18. There is also no reason at all why two, or even three, tracks can't pass underneath a component, so long as the tracks don't touch each other or touch the component lead holes.

If you were worried about your foil pattern being untidy, now's your opportunity to change all that. Place a sheet of tracing paper over a sheet of graph paper and tape the two together along the edges.

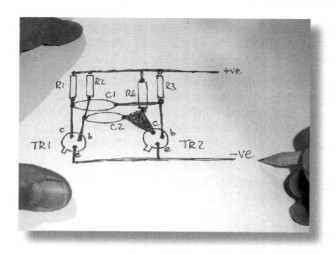

FIGURE 4.18   As the copper will be on the other side of the board to the components, there's nothing to stop you designing the printed circuit board so that copper tracks actually run beneath components.

Using the graph paper grid to align your components and connections accurately, redraw the layout on the tracing paper. This time, however, use two different colours of pencil, say, red for the component shapes and black for the connections. In this way you can instantly tell which is the 'top' and which is the 'bottom' of the board. The graph paper allows you to accurately get the sizes of components right, and position them as they will be on the final printed circuit board. Check, as you redraw the foil pattern, that it is correct with regard to the circuit diagram, and check you've remembered all components and got their polarities right.

And that's it! You've designed your own foil pattern.

You should now detach the tracing paper from the graph paper and keep it safely.

Pattern pending
This is the section which tells you how to transfer your pattern onto the copper-clad board, ready for etching.

As we saw before, the foil pattern must be put onto the board using etch-resistant materials. The form of this resist may be one of many: the simplest (and definitely the cheapest) can be ordinary household enamel or gloss paint, applied with a fine brush (messy); the most complex is photographic resist, developed on the circuit board much like your summer holiday piccies are developed on film and photographic paper (assuming you've not got a digital camera, of course). The one you use will depend on how much you're prepared to pay.

You can expect to pay about (gulp) £100 for a basic photographic system; this will let you make really ace printed circuit boards, but it won't do your pocket a lot of good — it's the sort of system which is likely to be used by schools and colleges where large numbers of one-off printed circuit boards are needed, so ask there first to see if they wouldn't mind you using theirs occasionally.

If you haven't got that sort of free cash around and can't find anyone willing to let you use their photographic system (ooh, the meanies) then you'll have to make do with a cheaper method.

Let's run through some of the methods now, so at least you know what the choice is.

Paint

The cheapest method of transferring your foil pattern onto the board is to use paint and a fine paint brush to hand-paint all the tracks and component holes. This is extremely messy, however, but will work quite well indeed, if a little care is taken.

A variant of this method is probably better though, where clear self-adhesive plastic has the image of the track pattern cut out of it (before the paper backing has been peeled off). The straight tracks can be cut out first, using a sharp craft knife and straight-edge, then the component holes can be punched out with a punching tool.

When the foil pattern is complete, you can peel off the paper backing and stick the plastic to the copper-clad board.

Now you effectively have a stencil, exactly the shape of your required foil pattern. There's no need to use a fine paint brush now; just paint over the whole plastic sheet with a much broader brush. After you've done that, carefully peel off the plastic and you're left with the foil pattern of painted resist.

Clever, eh?

Ink

Pens are available (ask at your local electronics store) which are loaded with special ink, resistant to the circuit board etchant.

With these pens it is quite an easy job (shown in Figure 4.19) to draw
the foil pattern directly onto the copper-clad board. You won't make a
perfectly neat printed circuit board using this method, but it does allow
a quick and easy route to a semi-acceptable finished job. Final quality,
however, depends totally on your artistic talent (or, at least, how steady
your hand is!).

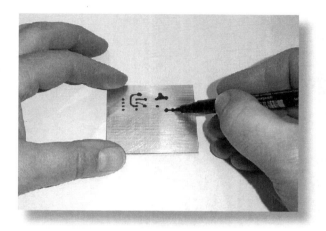

FIGURE 4.19   Drawing copper foil pattern directly onto the copper surface of a copper-clad board
using an etch-resistant ink pen.

Hint

In all these manual methods of applying resist it's sometimes a bit tricky
to estimate where on the copper board the component leads go, before
filling in the connecting etch-resistant bits. A useful trick is to put your
tracing paper copy of the foil pattern upside-down onto the copper (the
pencilled connections are to go on the bottom of the board, remember),
and using a sharp pointed instrument such as a scriber or a compass
point, prick through the tracing paper into the copper at every
component hole location.

When you remove the tracing paper you will find a network of little
holes, corresponding to each component lead hole.

Now you can put on the resist (whichever method you choose) knowing
where the positions are.

Transfers

Probably the best method of printed circuit board manufacture for the amateur is shown in Figure 4.20, where you can see that rub-down transfers are being used as resist, for component holes. You can buy transfers for all kinds of shaped holes, too — for integrated circuits, or edge connectors, or even specially shaped ones for transistors.

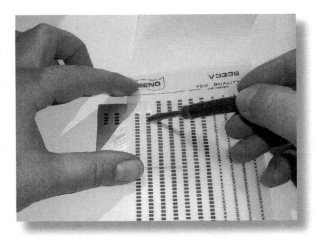

FIGURE 4.20   Rub-down transfers of all component shapes and sizes make an effective method of applying etch-resist.

When all the component locations are done, the connections can be made using straight-lined, or curved transfers, until the resist on the copper has the same shape as your foil pattern drawing.

Hint

A word of caution here: be extremely careful as you do this job, because any small cracks or breaks in the transfer resist will mean that a corresponding crack or break will occur in the copper when the board is etched.

FYI ◁

You don't have to stick to just one manual method of resist application. It's perfectly possible to combine two or more of these methods to good effect. For instance, you can use rub-down transfers to apply the component lead positions (particularly useful for integrated circuit positions) but use a resist pen, or paintbrush and paint, to make the other connections.

> **Hint**
>
> If you make any mistakes at this stage they can be removed using the corner of a small piece of sticky tape — but once you have etched the board your mistakes are permanent.

One minor variant also in this method is to substitute reels of special, thin, self-adhesive tape for these straight-lined or curved rub-down transfers, as the tape can be stuck to the copper surface in either straight lines or formed quite easily around curves as you stick it down. For obvious reasons, this is often referred to as the 'dots-and-tape' method of making printed circuit boards.

Photographic method

Using the photographic method, the highest quality printed circuit boards can be made.

What's more, as it *is* a photographic method, you can make more than one printed circuit board from it (just as you can get reprints of your holiday snaps).

Firstly you need a full-sized master copy of the foil pattern, on film. There are two methods here: where the master copy is a negative (as from a camera film); or where it is a positive. Both methods are used in printed circuit board making, but the positive method is the most popular, for no other reason than because it's easier to make a positive master than a negative one, so we'll stick with that one.

> **Hint:**
>
> You can make a positive master from your foil pattern by laying a sheet of acrylic film (which you can get from drawing-office suppliers) over the tracing paper or magazine foil pattern.
>
> Now, using the rub-down transfer or dots and tape method described earlier, make up the foil pattern directly onto the acrylic film. When complete, this film is your master copy — so keep it safe!

▷ **FYI** ◁

If your foil pattern is taken from a magazine or book, and you have a computer with a scanner and printer, you can make your master copy a little more accurately. Simply scan in the magazine foil pattern, then print it onto some acetate film. You'll need to check that the output size is the same as the source foil pattern and adjust as necessary. Note that some printers aren't capable of printing onto acetate film, however. Later there is more information about using computers to aid electronics construction.

Hint

If you are creating your own printed circuit board foil patterns for your own circuits, and you have a computer, then it might be worth looking at one or more of the many available programs for printed circuit layout. While these aren't always the easiest programs to use, they can give very good results, and if you intend making several printed circuit boards over the years, they are often worth the learning curve you have to climb. See later for details.

Resist-coated copper-clad board

The copper-clad circuit board you need now is not just ordinary board, but has a thin layer of *photographic resist* (sometimes shortened to *photo-resist*) over the copper, as illustrated in Figure 4.21, which — just like camera film — needs to be exposed to light. You can't just expose it to any old light, as you do with your camera film, though.

The only type of light which will affect the photo-resist is ultra-violet (UV) light.

Special light boxes (one is shown in Figure 4.22) are obtainable (it's these light boxes which cause this method to be expensive — they cost upwards of £80) in which the photo-resist coated board and the master foil pattern are placed, to expose the board, as illustrated in Figure 4.23.

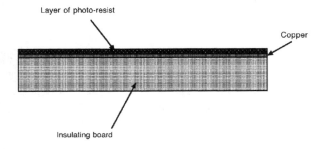

Layer of photo-resist

Copper

Insulating board

FɪɢᴜʀE **4.21** Photo-resist coated copper-clad board.

▷ **FYI** ◁

UV light is harmful to the human eye, in quantity, so these light boxes typically have a lightproof seal, although the power of radiation emitted by the internal UV tubes is not high anyway.

FIGURE 4.22   An ultra-violet light box.

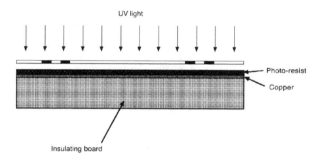

FIGURE 4.23   Exposing the photo-resist coated copper-clad board.

You must make sure that your foil pattern is the right way up here, because if it isn't your printed circuit board's track will be 'upside-down' or mirror-image, and the printed circuit board will be unusable!

After exposing the board through the master foil pattern, for the correct time (which depends on the strength of the UV tubes and the thickness of the layer of photo-resist, but is usually around five minutes or so), the board must now be developed.

The actual developing chemical depends on the photo-resist but is often a water solution of sodium hydroxide, and the exposed board is simply washed in a bath of it, as shown in Figure 4.24. As shown, the developing solution can easily be contained in a photographic-type developing tray, as it's not particularly hazardous.

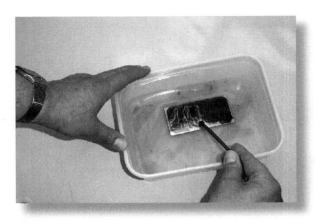

FIGURE 4.24   Developing an exposed, photo-resist coated, copper-clad board.

During development, the exposed areas of the board ie, the parts where no component holes or track connections were, soften and wash away. Meanwhile, the non-exposed areas corresponding to the foil pattern itself harden and remain in place, leaving the copper-clad board, complete with printed foil pattern.

Hint:

At this point in making a printed circuit board, it's always useful to check the foil pattern (which should be totally covered in etch-resist) to make sure that no breaks in coverage occur. If so, use a printed circuit board pen to patch broken areas.

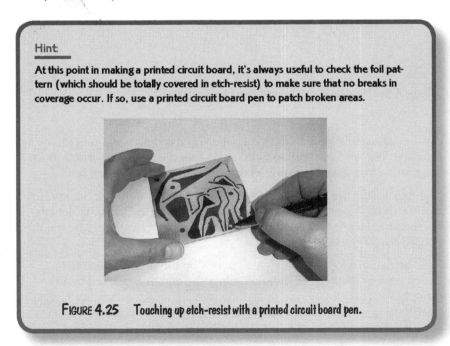

FIGURE 4.25   Touching up etch-resist with a printed circuit board pen.

If you wish to make more than one printed circuit board, you can use the acrylic film master foil pattern over and over again to expose other pieces of photo-resist coated board.

Using computers

The design of a printed circuit board copper foil track layout, and the later production of the printed circuit board from that copper foil are areas of electronics where accuracy is important. They are, indeed, almost sciences, rather than arts, although it can often help to have a steady hand.

As such, copper foil track layout is one particular area (there are others, some of which we'll see in this and other chapters) in electronics where computers can be put to good use, and there are several programs that can do this for us. Many of these programs form part of expensive industrial packages that take control of electronic device manufacture from initial marketing brief to late production stages. These packages are, needless to say, very expensive, and far more complex than we need to consider in this book.

On the other hand, there are several programs that are considerably cheaper than this, and even some that cost nothing at all to individuals.

Figure 4.26 shows a program called Crocodile Technology, which is an electronics simulator that allows us to build theoretical electronic circuits up on computer screen, testing them for function as required.

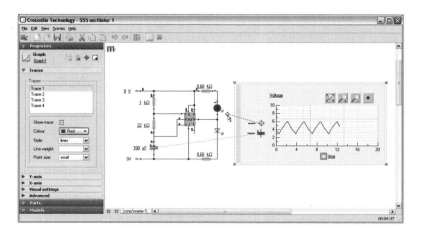

FIGURE 4.26   Crocodile Technology – an electronics simulator (available for Windows and Mac OS) that allows us to develop and construct circuits, on computer screen (Crocodile Clips).

By itself, Crocodile Technology doesn't aid us in producing copper foil track layouts for a printed circuit board, until it is used with a sister computer program called realPCB, which does just that — producing printed circuit board copper foil track layouts from the simulated circuits within Crocodile Technology. Figure 4.27 shows a portion of a printed circuit board layout produced using realPCB.

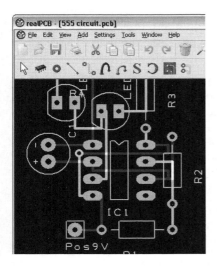

FIGURE 4.27   Part of a printed circuit board layout created using realPCB (Crocodile Clips).

While Crocodile Technology is available for Mac OS computer users, realPCB isn't available in Mac OS format. However, there are many alternatives for Mac users, one of which — Osmond — is a free program ideally suited for printed circuit board copper foil track layout.

Figure 4.28 shows a printed circuit board produced using Osmond — actually, the Bench Power Supply project's printed circuit board, in Chapter 8.

Programs such as realPCB and Osmond feature libraries of components, so that creating a printed circuit board layout is really just a matter of defining which components you want to use, then effectively joining the dots.

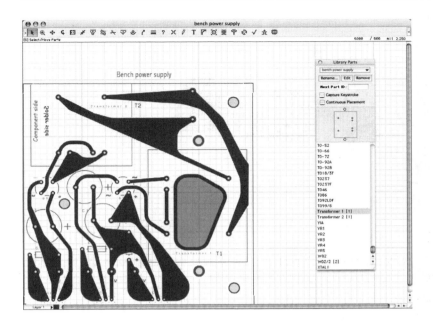

FIGURE 4.28   The Bench Power Supply copper foil track layout (see Chapter 8 for project details), produced using Osmond.

From screen to board

Producing a copper foil track layout on computer is one thing — but using that layout to make the actual printed circuit board is another! However, it is another area where computers can be helpful. As the computer can output to a printer, and as any printed circuit board layout program worth its salt can print its design, there are several ways that the computer can help us transfer the layout design onto the copper of copper-clad board, ready for etching.

Photographically

If you intend to use the photographic method detailed earlier, then it's a relatively simple process of printing the computerized copper foil track layout to film, ready for combination of the film with photographic resist-coated printed circuit board and an ultra-violet light box.

Purpose-produced translucent acetate sheet films are available for both laser and ink-jet printers (LaserStar for laser printers, JetStar for ink-jet printers). Figure 4.29 shows such a film that has been printed directly from a computer to a laser printer.

FIGURE 4.29   A printed circuit board design printed using a laser printer, on LaserStar translucent acetate sheet. JetStar sheets are used with ink-jet printers.

These translucent sheets are transparent to ultra-violet light (so are ideal for the purpose of ultra-violet light box exposure), but allow solid and dense areas representing the copper foil. Figure 4.30 shows the printed circuit board design of Figure 4.29 being used with an ultra-violet light box and photographic resist-coated printed circuit board.

FIGURE 4.30   Placing the LaserStar film and photographic resist-coated printed circuit board into an ultra-violet light box, ready for exposure.

Printed circuit boards of the highest quality can be made using this method, direct from a computer.

Non-photographically

If you don't have access to the ultra-violet light box photographic method of making printed circuit boards, there is an excellent alternative. Like the photographic method, it comes in the form of a specially-formulated acetate film that can be used with a printer (there is, though, only one type, that must be used only with a laser printer — it will not work with an ink-jet printer). The method is called Press-n-Peel, for reasons which will become clear.

The system is very easy to use, albeit a little tricky, and may take a few practice runs to get right.

First, print the printed circuit board copper foil track layout design, from whichever computer program you use, onto a Press-n-Peel sheet in a laser printer. Next, invert the sheet and place it onto the copper side of a piece of ordinary (ie, non photographic resist-coated) copper-clad printed circuit board, as shown in Figure 4.31.

FIGURE 4.31   Lay the Press-n-Peel sheet over the plain copper surface of the copper-clad board.

Next, lay a piece of plain paper over the top. Using a hot iron, press onto the paper to transfer the copper foil track layout image from the Press-n-Peel sheet onto the printed circuit board copper surface, as shown in Figure 4.32.

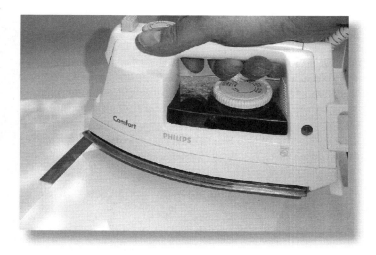

FIGURE **4.32**   Using an iron to transfer the image from a printed Press-n-Peel sheet onto the copper surface of plain copper-clad printed circuit board.

The iron must be hot (275–325°F) and you must keep it moving at all times to prevent the sheet melting.

After a short while, remove the iron and paper, then cool the printed circuit board and Press-n-Peel sheet under running, cold water. When the board is cold, peel off the Press-n-Peel sheet (Figure 4.33).

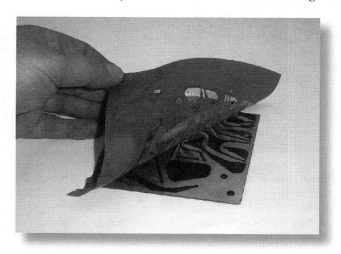

FIGURE **4.33**   Peeling off the Press-n-Peel sheet, following ironing and cooling.

Come up and see my etchings...

Finally, it only remains (did that man say *only?*) to etch the board and it's then ready for drilling and component insertion.

The etchant used to etch printed circuit boards is not actually an acid as you might expect, but ferric chloride — a horribly messy fluid — but it does its job well. Most local electronic stores will sell it bagged as crystals or granules and you simply mix them with water to give an etchant solution. A note of caution before you touch this stuff however: it is a very dangerous fluid and a strong solution of it will merrily eat its way through your clothes, the carpet, or even the stainless steel kitchen sink. So, please be careful and don't say I didn't warn you.

The etching process takes place more rapidly when the ferric chloride solution is warm and constantly moving, and for this reason you can buy specially built containers called *etching baths* to hold the solution. These etching baths are fitted with heaters to maintain a high temperature, and with agitators to keep the solution moving (vital to stop the build-up of copper oxide on the surface of the copper, which prevents further etching). Some top-of-the-range etching baths use a spray technique to maintain a high level of agitation — an etching machine of this type (which can actually be used with developer, too) is shown in Figure 4.34.

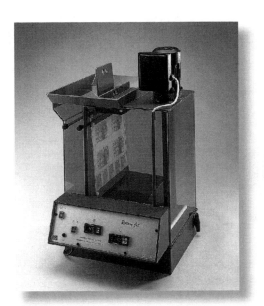

 FYI ◁

While historically ferric chloride has always been the etchant of choice in printed circuit board making, there are alternatives. While being somewhat less corrosive and nasty to use, these alternatives do tend to be a lot more expensive than ferric chloride however.

Figure 4.34   The Rota-Spray etching/developing machine (Mega Electronics UK).

After etching a number of printed circuit boards, the etchant will become exhausted and require renewal. You will notice this as a gradual increase in the time it takes to etch. When new, etchant may take only five or six minutes to etch a printed circuit board (particularly if the etchant is heated), whereas at the end of its working life etchant may take twenty minutes or more to do the job. The number of printed circuit boards a batch of etchant will successfully etch depends on the size of the printed circuit board and the area of copper foil to be etched, but you should realistically expect to be able to etch twenty or more printed circuit boards per batch.

> **Hint**
>
> Etching baths complete with heaters and agitators are (you guessed it!) expensive, but there's nothing to stop an enterprising reader making one, from nothing more complicated than a plastic container, aquarium heater, and a toy motor with plastic propellor. Just make sure that nothing metal is inserted into the etchant – for obvious reasons!

At the other end of the scale, a plastic photographic-type developing tray or similar can be adequately used as a makeshift etching bath, to hold the ferric chloride during etching, as shown in Figure 4.35. But as you'll have to keep rocking the bath gently, to agitate the etchant over the printed circuit board, this method is fraught with hazards — it's all too easy to splash some of the etchant.

FIGURE 4.35   A makeshift etchant bath – an old ice-cream container, in which the etchant is manually agitated by rocking the tray gently. A good excuse to have some ice-cream!

However you choose to do this job, it's wise to wear old clothes or overalls, and do the job outside, or at least in an uncarpeted workshop. Also, wear rubber gloves because it can do some pretty ghastly things to fingers — it'll certainly stop anyone's nail-biting problem, as it'll leave no nails at all to bite!

After etching, wash the board thoroughly in water to get rid of all traces of the etchant. Also, make sure that any etching solution has been either washed away or properly returned to its lidded container, and is not left where anyone (or even your cat, for that matter) can accidentally spill it.

Drilling for oil

Now is the point where all the component holes in the printed circuit board should be drilled so that components can be inserted and soldered. You can do this with a hand drill, or even a standard electric drill, but neither of these are particularly accurate and so may be the cause of irregular or misshapen holes.

A far better type of drill is the hand-held electric rotary tool commonly found in DIY and hardware stores. Being of quite high speed and easy to use they can make an excellent job.

Follow standard drilling practice when drilling a printed circuit board — put a block of wood under the printed circuit board as shown in Figure 4.36, so that you drill through the printed circuit board into the wood. This helps prevent the drill tip breaking roughly through the printed circuit board, and also helps prevent any damage to the worksurface.

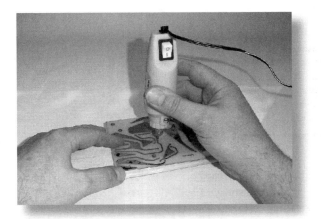

▷ **FYI** ◁

Make sure the drill bit (of whatever size you use) is perfectly sharp. A blunt drill increases drilling time, overworks your hand-held drill, and creates holes with burrs.

Figure 4.36   Hand-drilling a printed circuit board — rest the printed circuit board to be drilled on a block of wood. Hold the drill upright.

Irresistible

At this point, the etched and drilled board still has a coating of resist over the copper tracks of the foil pattern. Depending on your method of making the printed circuit board, you need to remove this resist prior to inserting the components and soldering. Resist removal can be done chemically using a solvent as shown in Figure 4.37.

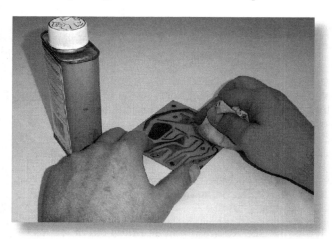

FIGURE 4.37   Removing etch-resist with a chemical.

> Hint:
>
> Some types of photo-resist coated boards actually allow soldering direct to the resist – in fact, the resist acts as a solder flux. If the resist-coated board you buy has such a resist, there is no requirement – other than personal preference – to remove it.

Such solvents are not cheap, so you can choose to remove the resist mechanically using an abrasive removal block (Figure 4.38) which you rub over the copper track to take off the resist. An alternative to a removal block is the fibreglass propelling pencil-type of abrasive cleaner, shown in Figure 4.39.

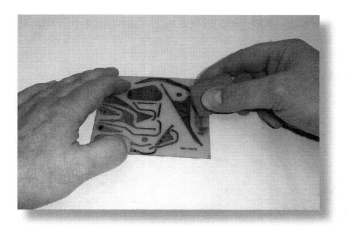

FIGURE 4.38   Removing resist using an abrasive block.

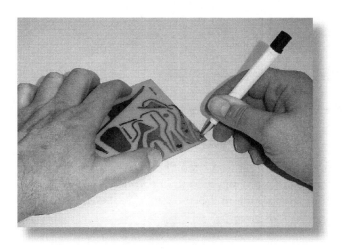

FIGURE 4.39   An abrasive cleaner in the form of a fibreglass propelling pencil-type of holder can also be used to remove resist.

Component insertion

As a rule of thumb, components should be inserted and soldered according to whether they can be damaged by heat. So, passive components like resistors and capacitors should be inserted and soldered first, leaving active components like integrated circuits until later.

Insert and solder only a few components at a time — no more than three or four at once for example. This means that components are less likely to slip out of their positions before solder is applied.

Once soldered into positions, trim all leads close to the soldered joint as shown in Figure 4.40. This prevents short circuits occurring between component leads.

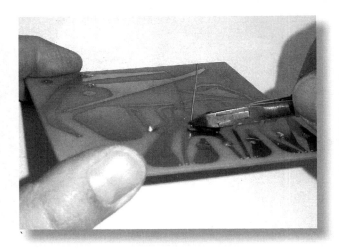

▷ **FYI** ◁

The process of inserting components into a printed circuit board is often — particularly in industrial electronics manufacturing — called (for fairly obvious reasons) stuffing.

FIGURE 4.40   Trim all component leads down after soldering. Use your side-cutters for this task.

Soldering

Soldering of components into a printed circuit board is actually quite an easy job, but there *is* a bit of a technique to it, and you need to be aware of the problems that incorrect soldering can create. Chapter 5 covers soldering in depth.

5

Soldering

The single most important process in electronics is soldering. It is — quite literally — the glue that holds all other processes together.

Soldering is the main means of:

- **HOLDING** components in their positions on the printed circuit board

- **CONNECTING** components to other components in the circuit.

What is solder?

Solder is a metallic compound which possesses a relatively low melting point. It has been used to join metals for thousands of years, since Roman times at least. Historically, one of the main constituents of solder has been lead. Indeed, the Latin name of lead — *plumbum* — gives a clue to one of solder's main uses, in plumbing, to join water pipes and fitments.

In electronics, solder's main use is to join component leads to printed circuit boards, in what is called a *soldered joint* — occurring between the component lead and the copper foil track of the printed circuit board. The copper foil track in turn allows that soldered joint to connect to other soldered joints (also formed by other component leads and the copper track), so that the complete electronic circuit can be made up on the printed circuit board. As a by-product, the rigidity of the soldered joints making up the circuit means that all components are also held securely in place.

The components of solder

Except in its very earliest uses, when pure lead was used, solder has always been a mixture of two or more substances. The Romans are known to have used a tin/lead alloy to joint lead water pipes. This is mainly because a tin/lead alloy has a lower melting point than lead itself. For example, pure lead has a melting point of 327°C, whereas an alloy of 62% tin and 38% lead has a melting point of 183°C — so it's possible to joint lead pipes, without actually melting the pipes themselves.

As soldering has evolved, and the uses to which it is put have changed, different solder alloys have been created for different purposes. In fact, soldering is still evolving — and is actually in the middle of quite a large evolutionary step — as we shall see here.

The 62% tin/38% lead solder alloy mentioned already, is known as the *eutectic composition*, and the melting point of 183°C (the *lowest* melting point of any tin/lead solder alloy) is correspondingly known as the *eutectic point*. It's not hard to understand that — purely in terms of soldering component leads to printed circuit board copper foil tracks in electronics — this has inevitably been viewed historically as a very useful, if not the ideal solder alloy.

However, times are changing. For a start, the world's resource of tin is depleting rapidly — only around 50 years' worth at current usage rates is left! More important though, is that lead is not exactly viewed as a safe substance, being a toxic, poisonous metal. For this toxicity reason, the world's electronics industry — along with other industries that currently use lead — is undergoing a revolution at present, to completely and permanently eliminate the use of lead in any of its processes and components.

Several new solder alloys have been developed to replace the old tin/lead alloys, in which various other elements — silver, bismuth, zinc, copper — have all been used with tin to form these new *lead-free* solder alloys.

While being inherently safer to both us *and* the environment, these lead-free solders all have a major disadvantage practically over the old tin/lead alloys in that the new alloys have higher melting points (around 30–40° higher!). This means that their very use may harm the electronic components being soldered, and they can be more difficult to use in general. Nevertheless, the electronics industry aims to be lead-free by 2006.

Effects of lead-free solder on us

The use of lead-free solder is mirrored in the non-industrial areas of electronics that readers of this book will be most interested in — in the home, school, college or university. Already lead-free solders are available off-the-shelf for personal use, and we can expect that tin/lead solders will be unavailable very much sooner rather than a little later.

The good soldered joint

Whatever the actual solder alloy used, the principles of making a soldered joint are the same. Solder is applied to a heated area of metal, whereupon the solder alloy melts and *wets* the metal to form a joint. Note that the term *wet* is used, and in some respects means much the same as a liquid such as water wetting a surface — it flows over the surface.

Wetting in soldering terms, on the other hand — unlike water wetting a surface — is a process where the solder comes into direct metallic contact with the metals to be soldered together into the joint. At the junction between the solder and the metal, a specific compound of both the solder and the metal is formed — known as an *intermetallic compound* — and occurs in a very thin layer between the two. Sometimes, the bond at that point is called the *intermetallic bond*. Interestingly, once this intermetallic compound between solder and metal has been created, it cannot be removed — you cannot remove solder from metal once it has been wetted!

Figure 5.1 shows a cross-section of a soldered joint, illustrating the main details. Figure 5.2 shows a micro-section of an actual soldered joint, in which can be seen the solder, the metal — copper — and the intermetallic compound layer between the two.

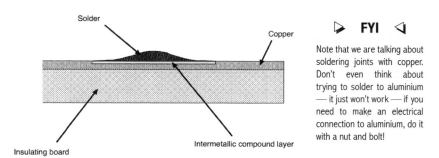

▷ **FYI** ◁

Note that we are talking about soldering joints with copper. Don't even think about trying to solder to aluminium — it just won't work — if you need to make an electrical connection to aluminium, do it with a nut and bolt!

FIGURE 5.1   Wetting takes place when solder comes into intimate contact with the surface metal, to form an intermetallic compound at the junction between solder and metal.

F**IGURE** 5.2   Micro-section through an actual soldered joint, where the intermetallic compound layer between solder on top and copper beneath is clearly visible.

The bad soldered joint

The condition for a good soldered joint to take place — illustrated previously in Figure 5.1 and shown photographically in Figure 5.2 — is that solder comes into contact with metal to wet it. Anything that *prevents* solder from coming into contact with metal, therefore, can work to prevent a good soldered joint from taking place.

The converse of the good soldered joint is one where the solder does not wet the joint, so does not come into contact with the metal. Such a bad soldered joint is shown in Figure 5.3.

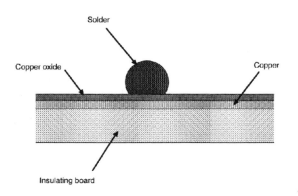

> FYI <

Many conditions can occur that create a bad soldered joint, and there are many types of bad soldered joints, so it is not always easy to specify exactly what has gone wrong in the soldering process. However, the usual problem is that the metal surface is not clean. Any number of contaminants may prevent a good soldered joint from taking place — dirt, grease, metal oxide, being the main ones.

F**IGURE** 5.3   A bad soldered joint – where a layer of metal oxide on the surface of the metal has prevented wetting of the solder to the metal taking place.

In practice, there are several levels of wetting which do not make a good joint, and these are worth considering as they give a clue as to the problems that we can experience in soldering electronic circuits.

All the various levels of wetting are described underneath, and Figure 5.4 shows the main levels by way of visual comparison.

- **NON-WETTING.** Pretty obvious to the eye — solder does not flow on to the metal at all. It will not adhere to the printed circuit board's copper foil track at all. A circuit built with no-wet joints will probably not work at all

- **PARTIAL WETTING.** While not flowing and completely wetting the metal surface, a partially wet joint will appear to be solid. However, it is unreliable, and may easily break off with time or vibration. Interestingly, the circuit itself may work initially when partially wet joints are present, but may not work for more than a short time

- **DE-WETTING.** Such joints may appear to have been created properly, but in the cooling phase after heat has been removed, the solder retracts — often leaving small solder balls around the joint. As in partial wetting, the joint may initially cause a circuit to work, but will break down fairly quickly

- **TOTAL WETTING.** Gives good joints, of course, which will create circuits that work and continue to do so. Note the difference between totally wet joints, and all the other wetting variants. A totally wet joint is concave in appearance (and will actually be quite shiny to the naked eye). All other wetting variants of joints will be convex (if not rounded) and may be dull in appearance to the naked eye.

Hint

As you can see here, one of the main causes (if not the main cause) of faults in electronic circuits, is bad soldering. Therefore, it makes sense to be able to solder well. Practise it before you attempt to build an electronic circuit till you are able to make good joints every time. Don't bother using printed circuit boards for this, instead use a few bits of stripboard (see page 81 for details) — even just a couple of boards give you many holes and copper strips to practise with. By the time you've soldered a couple of handfuls of components and a few short lengths of lead to the stripboards, you'll be a past master at the art of soldering!

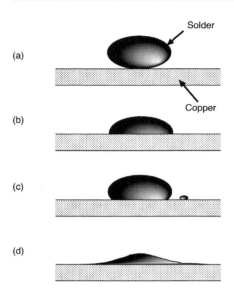

F<small>IGURE</small> **5.4** Four extremes of soldered joint: (a) non-wetting; (b) partial wetting; (c) de-wetting; (d) total wetting.

Flux

As we've seen, it's essential that the metal to be soldered is clean. The problem is that copper, the metal used to make copper-clad printed circuit boards that we use in electronics, oxidizes. A minutely thin surface layer of copper oxide forms on top of the copper-clad layer of the printed circuit board that makes up the copper foil track pattern very rapidly. So no matter how we clean the copper surface, the oxide layer builds up before we can get a soldering iron to it. And copper oxide cannot be soldered!

Further, every time a printed circuit board is handled it picks up dirt and grease, both of which work to prevent solder from wetting the board's copper track.

The key to making sure that the copper is returned to pristine condition in order that solder can wet it is a substance called *flux*.

Fluxes are chemically active — that is, they work on the surface contaminants (copper oxide, dirt, grease) on the copper track of the printed circuit board to dissolve them. This occurs when the copper is heated by a soldering iron, in the first stage of the soldering process. Once the copper surface is clean, solder can be applied.

▷ **FYI** ◁

There are many types of flux, judged usually by the level of chemical activity they have. Generally, the fluxes used in the solder used for hand soldering are of low activity. They are typically made by distilling the sap (or resin) from pine trees. The residue is then dissolved in a solvent, making it easier to apply.

At first sight, it might seem that co-ordinating all this (heating the copper, applying the flux, waiting till the copper is clean, applying the solder) is a tricky matter, best left to the professionals and their industrial soldering processes. However, the *real* trick is that it can all be done by hand in a single smooth process by anyone, as long as a few simple steps are followed.

Key to the whole process is that modern solder, used to solder printed circuit boards by hand, already contains the flux needed to clean each joint to be soldered.

The solder used has cores (usually four or six) throughout its length which are filled with flux — as illustrated in Figure 5.5. Diameter of the whole flux-cored solder lead is usually around 1 to 2 mm.

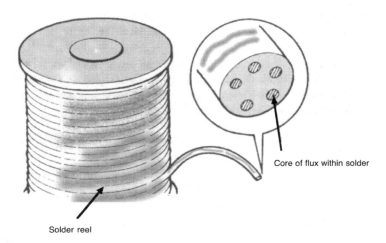

Core of flux within solder

Solder reel

FIGURE 5.5   Cored solder, obtained in reels, appearing much like thin flexible wire. Flux is present in cores within the solder.

So how does it work? After all, having the flux inside the solder like this means that the flux is presented to the joint to be soldered at the same time as the solder is, rather than before it — which would suggest it has no time to clean the joint. However, the flux has a much lower melting point than the solder has, so the flux melts and flows onto the joint first. It therefore has the time to coat the metal surfaces of the joint and clean them, before the solder melts and flows over the joint to make the soldered joint, as shown in Figure 5.6.

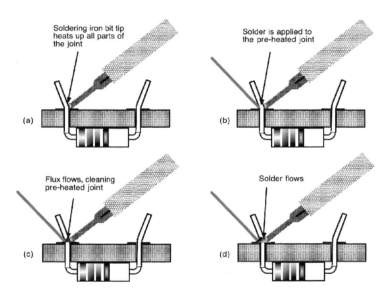

Soldering iron bit tip heats up all parts of the joint

(a)

Solder is applied to the pre-heated joint

(b)

Flux flows, cleaning pre-heated joint

(c)

Solder flows

(d)

FIGURE 5.6 The main stages of hand soldering a joint: (a) the soldering iron is applied to the joint, to heat it; (b) while maintaining the soldering iron's position, cored solder is applied to the joint; (c) flux melts quickly and flows from the cored solder onto the joint's metal surfaces, to clean them and protect them; (d) the solder melts, and flows over the joint's metal surfaces.

▷ **FYI** ◁

This whole process, from first applying the soldering iron to the solder flowing over the joint surfaces, takes only a few seconds (depending mainly on the size of the joint and how powerful the soldering iron is). If a joint takes longer than just a few seconds to make, there's a good chance the soldering iron isn't powerful enough.

Take Note

Throughout this chapter we look at processes involved in soldering various joints on printed circuit boards (ie, those circuit boards made from copper-clad board, with copper foil tracks). The processes – however – are actually identical whatever sort of circuit boards you use.

Key points when soldering

Being able to use a soldering iron isn't the only thing you need to know when soldering. The whole process depends on several aspects:

When soldering

- **CLEAN** all parts — copper printed circuit board track, component leads, soldering iron bit — *before* soldering

- **MAKE** a reliable mechanical joint — *before* soldering

- **HEAT** the joint sufficiently — *before* applying solder

- **APPLY** the solder — keeping the soldering iron on the joint while the solder melts

- **REMOVE** the soldering iron and allow the solder to solidify — *before* handling or moving the joint

- **TRIM** the excess component lead from the joint, once the joint has solidified and cooled sufficiently to allow handling.

We should now consider these six aspects in detail.

Cleaning all parts

We already know that flux is used automatically when soldering using cored solder to help clean the joint surfaces, as part of the soldering process. However, flux can only remove so much oxide, grease or dirt. It's best to improve your chances of making a good solder joint by manually cleaning the joint surfaces first. Use an abrasive fibreglass pencil (as shown in Figure 5.7), or an abrasive scrubbing block (as shown in Figure 5.8), rubbing them over both copper track and component leads to minimize dirt and grease.

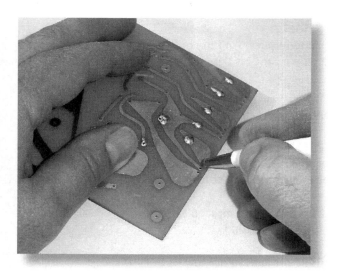

FIGURE 5.7   Cleaning copper track of a printed circuit board, using an abrasive fibreglass pencil – simply rub the pencil over the track, particularly over the component land areas to clean it. Brush aside the fibreglass residues before commencing soldering.

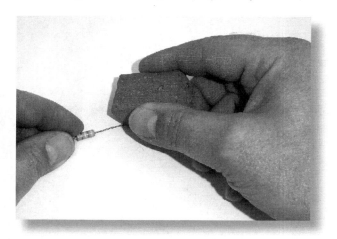

FIGURE 5.8   Using an abrasive scrubbing block to make sure a component's leads are clean, prior to soldering.

It's vitally important that the soldering iron bit tip be kept clean too, and the best way to do this is to *tin* it. Tinning is a process whereby fresh solder is regularly melted on to the heated bit tip, so that the tip remains coated with fresh solder. As flux-cored solder should be used, just as when soldering the actual joint, the flux within the cores cleans the tip simultaneously.

The process is a two-stage one: first wipe old excess solder off the soldering iron bit tip, then tin the tip by applying fresh solder. Figure 5.9 shows a wet sponge being used to wipe the soldering iron bit tip on. An alternative is shown in Figure 5.10, where a tip cleaning and tinning block is being used.

> **Hint**
>
> If you have bought yourself a new soldering iron and are preparing to tin it, you will find that when you first plug the soldering in to turn it on, smoke will be created and given off from the bit tip: this is because the bit is coated with a grease substance to prevent it oxidizing. This is useful, but also means that as the bit heats and burns off the grease, oxide very quickly forms, making the tip unusable. To prevent this, tin the bit tip as shown in Figure 5.11 immediately as it heats.

FIGURE 5.9   Cleaning a soldering iron bit tip using a wet sponge. Simply wipe the tip over the sponge surface.

FIGURE 5.10   Using a tip cleaning and tinning block to clean a soldering iron bit tip — just wipe the tip over the block.

Immediately following removal of old excess solder, apply fresh solder to the soldering iron bit tip, as shown in Figure 5.11. Just a small amount of solder is all that's necessary to keep the tip in err... tip-top condition.

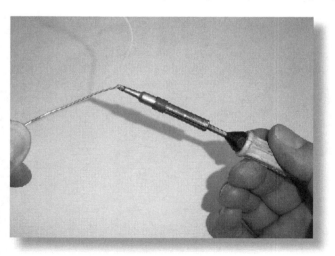

FIGURE 5.11   Tinning a soldering iron bit tip. Once the soldering iron bit tip has been cleaned as shown in either Figure 5.9 or Figure 5.10, quickly tin it using a small amount of ordinary flux-cored solder. The tip will have a shiny appearance if tinned properly.

Making a reliable joint

Components must not move while you are soldering a joint, otherwise
the joint may be faulty. As a general rule, it's best to fix the components
in place mechanically somehow before soldering. With axial- or radial-
leaded components such as resistors and capacitors, the easiest way to
do this is to bend the leads slightly after insertion into the printed circuit
board, as shown in Figure 5.12. As a guide, the angle of the bend doesn't
need to be (and *shouldn't* be) any more than just a few degrees — bend
the leads only as far as necessary to hold the component in place.

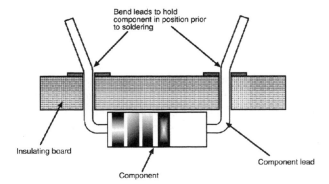

Bend leads to hold
component in position prior
to soldering

Insulating board

Component lead

Component

FIGURE 5.12   After inserting a component into the printed circuit board, bend its leads slightly to hold
it in place ready for soldering.

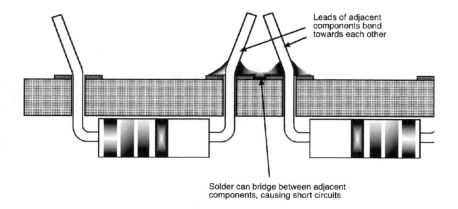

Leads of adjacent
components bend
towards each other

Solder can bridge between adjacent
components, causing short circuits

FIGURE 5.13   Showing how solder can easily bridge between joints if angled component leads are used to hold components in place prior to soldering — take care when using this method to prevent such short circuits.

Heat the joint

For successful soldering, the metals to form the soldered joint should be pre-heated. Pre-heating is easily accomplished — apply the soldering iron bit tip to the joint — touching both the copper track and the component lead, as shown in Figure 5.14. The time you should pre-heat the joint depends on the size of metal to be heated and the power of the soldering iron, but generally this will be for no more than a few seconds — between about two and eight seconds, say. You will learn to judge the pre-heat time required by joints with practice, but in any case it's not normally *that* crucial.

The factors you should bear in mind are that:

- **IF** you don't pre-heat the joint sufficiently, the molten solder cools too quickly and you run the risk of making a defective joint

- **IF** you pre-heat the joint too much, you may damage the component and may even cause the copper track to lift from the printed circuit board.

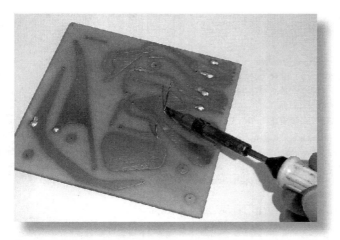

FIGURE 5.14   Pre-heating the joint to be soldered – apply the soldering iron bit tip to the joint's metal surfaces.

Apply the solder

With the soldering iron bit tip in position, apply the end of the flux-cored solder to the joint — touching it to both the joint and the soldering iron bit tip. If everything is as expected, the solder will melt and flow around the joint. Figure 5.15 shows this stage.

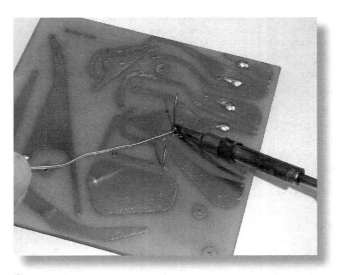

FIGURE 5.15   Solder melting and flowing around the pre-heated joint.

Remove the soldering iron

Once the solder has flowed around the joint (not before), quickly remove the soldering iron. Do not move the printed circuit board, and do not move the component or its leads, until the solder has solidified. If you move anything before the solder solidifies, the joint may be damaged.

▷ FYI ◁

A good soldered joint has a concave, shiny appearance. If the joint is not heated sufficiently, or if you move the joint and damage it, you will probably notice that it is not concave, or that it is a dull, grey colour. Don't worry — just repair the fault by heating the joint again, until the solder flows properly.

Trim excess component leads

If the component leads were left as they are, and more than just a couple of components were inserted into the printed circuit board, you run the imminent risk of component leads touching — effectively forming short circuits. To prevent this you have to trim all excess component leads off immediately above the joint. Figure 5.16 shows this.

After all joints on the printed circuit board have been soldered and excess leads trimmed, it's a good time to check all the joints visually to see if they look 'good' or 'bad', repairing any that are suspect. Also check over all the copper foil track, to see that no solder bridges have formed between tracks or component joints. If you find any solder bridges, use desoldering tools — a solder sucker, or desolder braid — to remove excess solder, and resolder any joints that need it.

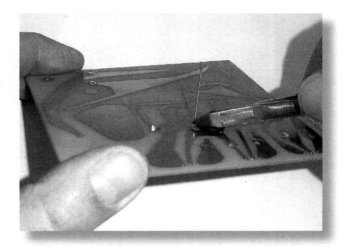

FIGURE 5.16   Trimming excess component leads off, close to the soldered joint, to prevent short circuits.

Connecting leads

Inevitably, any printed circuit board holding its electronic circuit is not isolated from the world around it! It must be connected — to switches, controls, off-board components, input and output sockets, and so on. So, it's common — indeed, normal — to need to solder connecting leads to the printed circuit board copper foil track.

To solder connecting leads to a printed circuit board, use the following procedure:

- **STRIP** around 5 mm of insulation from the end of the lead

- **IF** the connecting lead is multi-stranded — twist the loose ends of the stripped lead. If you solder the lead untwisted to the board, there's a possibility a loose strand may bridge across to another track

- **CLEAN** and tin the soldering iron bit tip. Yd   ALWAYSd
  before any soldering operation

CONTINUED...

...CONTINUED

- **TIN the connecting lead.** This has two purposes: first, it removes any oxides, dirt and grease on the wires making up the leads, and; second, the solder you add in the tinning stage will be used when the joint is made — meaning you won't have to add much solder at that time. The soldering iron bit tip needs to be quite a broad one — around 3 mm — for this job. To tin the lead:

▷ hold the flat part of the soldering iron bit tip upwards (as shown in Figure 5.17) and lay the stripped end of the connecting wire on the tip's flat surface

▷ apply the end of a length of solder to the wire (not the soldering iron bit tip) — this means you can see when the wire is hot enough, as the solder begins to melt

▷ when enough solder has flowed over the lead ends to fully cover the wire, remove the soldering iron and solder

▷ allow the lead to cool before touching or moving it.

FIGURE 5.17   Tinning a connecting lead – do not heat the lead for too long, as insulation may melt or deteriorate.

CONTINUED...

...CONTINUED

- **TIN the printed circuit board.** At the point where the tinned lead is to be connected to the printed circuit board's copper foil track, the copper must be tinned, to remove oxide, dirt and grease, and to apply solder ready for the joint to be made between board and lead. To tin the printed circuit board track:

  ▷ clean and tin the soldering iron bit tip

  ▷ apply the soldering iron bit tip to the copper track of the printed circuit board, at the point where the lead is to be connected — as shown in Figure 5.18 — and wait for a few seconds

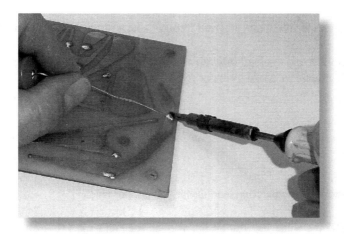

FIGURE 5.18   Tinning the copper foil track of a printed circuit board — apply the soldering iron bit tip first, and wait for a few seconds before applying solder.

  ▷ apply the end of the flux-cored solder to the track, a small distance (about 2–3 mm) away from the soldering iron bit tip — wait for the solder to melt and flow towards the soldering iron bit tip

  ▷ remove the soldering iron bit tip and the solder

  ▷ allow the printed circuit board to cool before moving or touching the
     b

> Take Note
>
> While it's correct to tin the printed circuit board copper foil track at the points where connecting leads are to be jointed, this is NOT the procedure where components are to be fitted to the printed circuit board — if the copper foil track is tinned where a component lead is to be inserted, the hole for the lead may be closed with solder.

Once the lead is tinned, it can be soldered to the copper foil track of a printed circuit board. There are three main ways this can be done, soldering the lead:

- **1** directly to the printed circuit board copper foil track

- **2** to a terminal pin, inserted through a hole in the copper foil track

- **3** inserting the lead through a hole, purposely designed into the printed circuit board's copper foil track.

The procedures you should follow for all three ways are detailed below.

1 Soldering a connecting lead directly to copper track

- **TIN** the soldering iron bit tip, as described on page 116

- **TIN** the connecting lead, as described on page 124

- **TIN** the printed circuit board copper track, at the point where the lead is to be connected, as described on page 125

- **HOLD** the tinned end of the lead on the tinned copper track, as shown in Figure 5.19

CONTINUED...

...CONTINUED

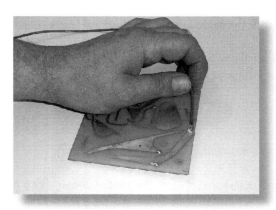

FIGURE 5.19   Position the tinned connecting lead end over the tinned copper track.

- **APPLY** the soldering iron bit tip to the tinned lead end (not the copper track), pressing down on the lead end to hold it in position, and wait for the solder on both the lead end and the copper track to melt, as shown in Figure 5.20

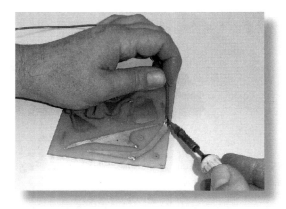

FIGURE 5.20   Heating the connecting lead end and copper track – the solder already present in the tinning on both the lead end and the track will be sufficient to make the soldered joint.

CONTINUED...

...CONTINUED

- **REMOVE** the soldering iron, but hold the lead without moving it

- **WAIT** for the solder to solidify before moving the printed circuit board or letting go the lead.

---

2 Soldering a connecting lead to a terminal pin

- **INSERT** a terminal pin into the printed circuit board — note that the pin goes through from the copper foil track side of the printed circuit board to the component side of the board ie, the connecting lead will be soldered to the component side NOT the copper track side (see Figure 5.21).

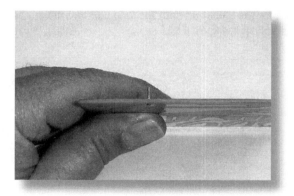

FIGURE 5.21   A terminal pin goes from the copper track side of a printed circuit board, to the component side.

- **CLEAN** and tin the soldering iron bit tip, as described on page 116

- **TIN** the connecting lead, as described on page 124

CONTINUED...

...CONTINUED

- **APPLY** the soldering iron bit tip to the terminal pin and wait for a few seconds (See Figure 5.22)

FIGURE 5.22   Soldering a terminal pin to a printed circuit board.

- **APPLY** solder to the terminal pin

- **WHEN** the solder melts and flows over the terminal pin and the copper track, remove the soldering iron and wait for the soldered joint to cool

- **TURN** the printed circuit board over, then tin the terminal pin

- **HOLD** the tinned end of the connecting lead against the tinned terminal pin, as shown in Figure 5.23

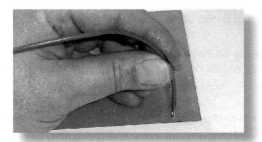

FIGURE 5.23   The connecting lead should be vertically aligned with the terminal pin, and touching it.

CONTINUED...

...CONTINUED

• **APPLY** the soldering iron bit tip to the connecting lead and terminal pin

• **WAIT** for a few seconds for the solder on both the lead and pin to melt, then remove the soldering iron

• **DO** not move either connecting lead or printed circuit board, until the solder has solidified.

Hint

For both these methods of soldering a connecting lead to a printed circuit board, be extra careful not to allow the connecting lead end to move at all once the joint has been soldered and during the cooling phase, or a defective joint may result.

3 Soldering a lead to a hole

• **CLEAN** and tin the soldering iron bit tip, as described on page 116

• **INSERT** the connecting lead through the hole in the printed circuit board — do not tin the connecting lead first, or it will not go through the hole!

• **APPLY** the soldering iron bit tip to the joint, touching both the copper foil track of the printed circuit board and the lead, as shown in Figure 5.24

CONTINUED...

...CONTINUED

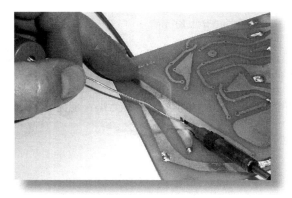

FIGURE 5.24   Soldering a connection lead to a printed circuit board.

• **APPLY** solder to the joint

• **WHEN** the solder melts and flows over the connection lead and the copper track, remove the soldering iron and wait for the soldered joint to cool

• **DO** not move either connecting lead or printed circuit board, until the solder has solidified.

Unsoldering

There are times when it's necessary to unsolder components from a printed circuit board. You may, perhaps, have inserted and soldered a component incorrectly into place, or a particular component may be faulty. In such cases, unsoldering is the only option.

There are two basic methods of unsoldering components (detailed overleaf), and both rely on having extra tools: either solder braid, or a desoldering pump.

Using solder braid

- **TIN** the soldering iron bit tip

- **PLACE** the end of the solder braid over the joint to be desoldered, as shown in Figure 5.25

- **APPLY** the soldering iron bit tip onto the solder braid. Press gently but firmly against the joint. After a few seconds or so, the solder will melt, and some will wick into the solder braid

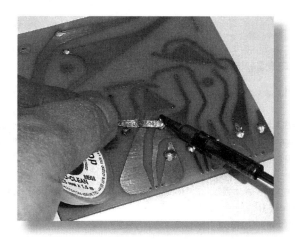

FIGURE 5.25   Desoldering using solder braid – the molten solder wicks into the strands of the solder braid away from the joint.

- **REMOVE** the soldering iron bit tip. Place a fresh portion of solder braid onto the solder joint and repeat the last step until all the solder forming the joint has been removed

- **REPEAT** the process for all component joints

- **REMOVE** the component.

Hint

As you use a portion of solder braid to unsolder a joint, it will become loaded with solder so cease to be usable. Using a pair of side-cutters, trim off the used portion to maintain a fresh solder braid end.

Hint

Sometimes not all solder can be removed from a joint (very often a small amount which cannot be removed will still hold the component lead to the copper track). If this occurs, you will find that you have to heat the joint with the soldering iron bit tip one last time, before quickly — before the solder solidifies — releasing the component.

Using a desoldering pump

- **TIN** the soldering iron bit tip

- **PRIME** the desoldering pump, by pressing the charger down, as shown in Figure 5.26

FIGURE 5.26   Priming the desolder pump by pressing down the charger.

- **APPLY** the soldering iron bit tip to the soldered joint and wait for the solder to melt

CONTINUED...

...CONTINUED

- **APPLY** the nozzle of the desoldering pump to the molten solder, as shown in Figure 5.27

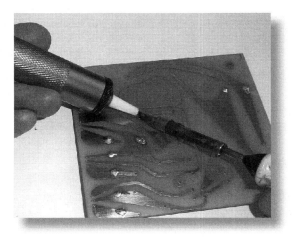

FIGURE 5.27   With the soldering iron bit tip heating the solder, apply the desoldering pump nozzle.

- **PUSH** the pump's release button. The vacuum will suck the molten solder away from the joint and into the desoldering pump chamber, where it will solidify

- **REPEAT** the previous four steps as required, until all the solder has been removed from the joint.

Take Note

Every now and again, the desoldering pump chamber will fill with solidified solder to the point that it doesn't work efficiently. Empty it by repeatedly priming and discharging the pump, so that the solder is forced out of the nozzle.

> **Take Note**
>
> Unsoldering usually requires considerably more application of heat than soldering – the soldering iron bit tip is applied to the joint more times and for longer periods than when soldering.
>
> Remember that components may be damaged by excessive heat, so space the desoldering operations out to allow the components to cool down in between.

Care of your soldering iron

Your soldering iron is a tool — a vital one when it comes to building electronic circuits, of course — and like any tool needs looking after, to keep it in good working order throughout its potentially long life. Things are made a little complicated in that the soldering iron is — I'll state the obvious here — very hot, so you also need to take certain safety precautions while using it. But, fortunately, on the other hand it's a fairly simple tool, so the things you need to do to look after it and the precautions you need to take while you're using it aren't excessive.

Looking after your soldering iron

- **USE a stand.** When at rest, a soldering iron can be easily knocked or moved — it may fall off the work surface or simply touch against something else. Use a stand (as shown in Figure 5.28) when you are not actually soldering with it

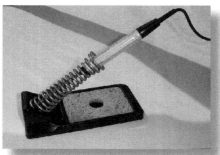

FIGURE 5.28   Use a soldering iron stand when not actually using your soldering iron. Keep the safety risks to a minimum.

CONTINUED...

...CONTINUED

- **TIN** the soldering iron bit tip regularly — even at rest. When you solder with a soldering iron it is regularly in contact with fresh flux-cored solder, so the bit tip is maintained in good condition. On the other hand, when you are not actually soldering and a soldering iron is standing unused but still turned on, the bit tip can become depleted of solder and flux due to the heat, so it oxidizes and become unusable. If you are not going to solder with the soldering iron for a while, remember still to tin it every few minutes

- **TURN** the soldering iron off when unused. To reduce safety risks to an absolute minimum, and to help prevent the soldering iron bit tip from becoming unusable, turn the soldering iron off if you are not going to use it in the next ten minutes or so

- **CLEAN** the soldering iron bit tip when cold by lightly rubbing with a nylon pad. This may help a soldering iron bit tip that will not wet — but do not use wire wool or emery paper to clean the soldering iron bit tip, as this will remove protective plating and shorten the bit life

- **NEVER** put a soldering iron — hot or cold — into liquid

- **CHECK** the soldering iron's cable regularly for burns. A soldering iron when on is, of course, very hot, and accidental burning on the mains cable is a possibility.

Printed circuit board links

If a circuit is complex, it's probable that the printed board copper foil track layout will be complicated too. Under these circumstances, it's often the case that it's simply not possible to design a copper foil layout in a single layer. One trick to get around this problem is to use the component side of the board as well as the copper track side to complete the circuit connections. This is done with the use of simple links of wire, which connect from one point on the copper foil track, go over the track on the component side, then connect to another point on the copper foil track. Figure 5.29 shows an example of a printed circuit board which uses links.

FIGURE 5.29   A printed circuit board that uses links to aid in copper foil track circuit layout. Note that many copper foil connections can take place under links, thereby easing a layout considerably. This particular printed circuit board is the Heart Flasher project, in Chapter 8.

Making a link:

- **CUT a short length of single-strand wire.** This should be a little longer than the distance the link needs to span between holes on the printed circuit board

- **BEND one end of the link around the nose of a pair of long-nosed pliers — see Figure 5.30.** The bend should be at 90°, and there should be sufficient wire to go through the board and be soldered to a copper track

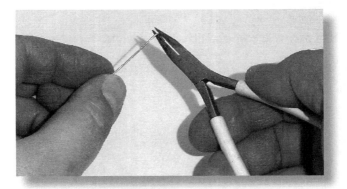

FIGURE 5.30   Bend one end of the link, using long-nosed pliers.

CONTINUED...

...CONTINUED

- **MEASURE** the link length required

- **BEND** the other end of the link around the long-nosed pliers
  — see **Figure 5.31.** The length of link between bends should be the same as that you've just measured

FIGURE 5.31   Bend the other end of the link.

- **INSERT** and solder the link

- **TRIM** excess link wire off the soldered joints.

6

Housings

After the printed circuit board, the next most important part of an electronic project is the *housing* (sometimes called an *enclosure* or *case*). OK, yes, while it's perfectly possible to use an electronic circuit just at its printed circuit board stage (complete with controls and connections hanging off…), it's certainly not the best method of ensuring your circuit has a long working life. Only the housing can 'finish' an electronic circuit off, and turn it into a complete design.

More than just the practical aspect of ensuring the internal electronic circuit is securely and safely held, however, in many respects the housing forms the main (if not the only) interface between the electronic circuit and the user. Yes, the electronic circuit may have connectors for inputs and outputs, and have controls that the user adjusts, but the housing is the means whereby all the controls and connectors are presented for use. How effectively this occurs depends very much on the housing, and how it has been designed. And there's a world of difference between a good housing and an acceptable one.

In fact, to emphasize how important an electronic circuit's housing is, we should look at a complete electronic circuit and see how it breaks down into a total of six levels of interconnection — as shown in Figure 6.1. In this look at the electronic circuit, we have to consider an integrated circuit as being a major component within the circuit itself, in order to understand the various levels and how they interact. Normally, of course, we would most likely use an integrated circuit, without understanding — or even caring — how it is constructed internally. This example, though, shows just how important each particular level is, with regard to the whole electronic circuit design.

The six levels are:

- **LEVEL 1 — on-the-device.** Where individual connections between the semiconductive component parts of integrated circuits are made as minute etched wires — shown in Figure 6.1(a)

- **LEVEL 2 — device-to-package.** Where the semiconductor die is permanently attached, with thin aluminium or gold wires, to the terminals of its package — shown in Figure 6.1(b)

- **LEVEL 3 — package-to-board.** Where flat, round or square package terminals are connected to printed circuit boards, as permanent (ie, soldered) or separable connections — shown in Figure 6.1(c)

- **LEVEL 4 — board-to-board and on-board.** Where components are interconnected with the conductive track of the printed circuit board, and where printed circuit boards are connected to each other by permanent or seperable connectors and cables — shown in Figure 6.1(d). The housing itself is directly involved at this level, simply because the types of connectors you use, and the printed circuit board positioning with respect to these connectors, may well be defined by the housing

- **LEVEL 5 — board-to-housing.** Where connections between a printed circuit board and its housing are made with permanent or separable connectors — shown in Figure 6.1(e). Sometimes, a distinct chassis may be involved between printed circuit board and housing

- **LEVEL 6 — housing-to-housing.** Where separable connections between housings are made with connecting cables — shown in Figure 6.1(f).

Figure 6.1 shows quite graphically what levels of an electronic circuit design the housing affects. Levels 1 to 3 (Figure 6.1(a) to (c)) have no involvement with the housing at all. They are merely concerned with connections between the individual components' innards and the printed circuit board. On the other hand, levels 4 to 6 (Figure 6.1(d) to (f)) are all indeed affected by the housing or housings you might choose for your electronic design.

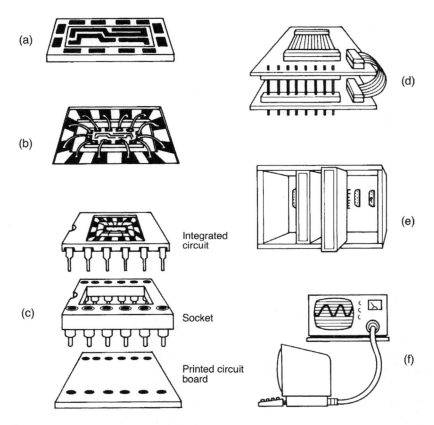

Figure 6.1 Six levels of interconnection between parts of an electronic design: (a) semiconductor die; (b) from the die to the terminals of the integrated circuit package; (c) from the integrated circuit to the printed circuit board; (d) between printed circuit boards; (e) between printed circuit boards and the housing; (f) between housings.

Yet, despite this fact that the housing affects or is affected by no less than half of the total electronic design, most magazines and books in general pay little attention to the housing, other than a few spartan words about putting the printed circuit board in to some sort of box. Indeed, only rarely is the actual housing design taken into consideration at all.

This chapter aims to put that right. Hopefully, by the end of reading it, you will see that the housing is as important a part of the total electronic device as the circuit design itself.

What's your housing made of?

There are three main types of housing available to the beginner in electronic construction:

- **PLASTIC**

- **METAL** diecast

- **METAL** folded chassis-type.

 ▷ **FYI** ◁

Actually there are more housing types, but these three types (and their derivatives) form the largest bulk of housings you will come across to begin with. As you progress in electronics you will undoubtedly meet other types.

Figure 6.2 shows examples of these housing types.

FIGURE 6.2   Housings – from left to right – plastic, metal diecast, metal folded chassis-type.

Plastic

Plastic housings come in several forms. They are relatively cheap to make in bulk, very easy to use, and provide a neat method of housing electronic circuits. Being usually moulded or extruded, some quite intricate and novel plastic housing shapes are available, which means that it's often possible to find a plastic housing that is a good match for an electronic circuit's function — for example, a plastic housing is shown in Figure 6.3 that is provided with a belt clip, and internal battery space, which would make an ideal housing for something like a music player or a pocket game.

FIGURE 6.3   A small plastic housing, complete with belt clip and battery compartment.

Metal diecast

Made from cast aluminium, diecast housings are extremely tough, yet quite easy to work with. They are not, on the other hand, the most attractive of housings, so probably wouldn't be your first choice to house an electronic circuit that you wanted to use, say, on display in your living room, but a diecast box would be great for, say, a small pre-amplifier for use between an old record deck and a modern hi-fi amplifier.

Metal folded chassis-type

Usually made from just two thin sheets of aluminium, each folded to complement, metal folded chassis-type housings make very effective enclosures for many electronic circuits. The bottom half of a housing is where a printed circuit board and controls are fitted, while the top half of the housing forms a neat lid, allowing full access when removed, and protection when fitted. Often, the top half of such housings are plastic coated, or painted, which gives quite an attractive appearance to the finished electronic device.

Other housings

There are some other housing types that you should be aware off too, notably those made from extruded aluminium. Such housings are very tough, and have internal slots which are useful for the mounting of printed circuit boards — boards just slide into position from one end of the housing.

IP rating

In certain environments such as the great outdoors, it's necessary to use a housing that protects the electronic circuit inside from the elements to some degree or other. While all housings will protect their innards to some small measure, the problem is obviously magnified where the electronic device is to be used in such possibly detrimental environments as outside in the rain. For many years, there has existed a method of determining whether an enclosure is suitably protected against the environments in which it's expected to be used. This is a standard method, called the *ingress protection* (*IP*) rating system. It comprises two or three digits, and each digit refers to a particular level of protection, for a particular type of protection requirement. Thus, an enclosure might be given an IP rating of, say, IP63, or another might have an IP rating of IP485. Table 6.1 gives the IP ratings and numbers.

Where only two digits are given in an enclosure's IP rating, the third of these requirements (impact damage) is not relevant so should be ignored.

TABLE 6.1   Ingress Protection (IP) ratings by number, and their equivalent protection levels.

Val-ue	1st digit	2nd digit	3rd digit
	Protection against ingress of solids	**Protection against ingress of liquids**	**Protection against mechanical impact damage**
0	No protection	No protection	No protection
1	Protected against solid objects over 50 mm eg, hands, large tools	Protected against vertically falling drops of water	Protected against 0.225 joule impact (150 g @ 15 cm)
2	Protected against solid objects over 12 mm eg, hands, large tools	Protected against direct sprays of water up to 15° from vertical	Protected against 0.375 joule impact (250 g @ 15 cm)
3	Protected against solid objects over 2.5 mm eg, wire, small tools	Protected against direct sprays of water up to 60° from vertical	Protected against 0.5 joule impact (250 g @ 20 cm)
4	Protected against solid objects over 1.0 mm eg, wires	Protected against water sprayed from any direction. Limited ingress permitted	
5	Limited protection against dust ingress (no harmful deposit)	Protected against low pressure water jets from any direction. Limited ingress permitted	Protected against 2.0 joule impact (500 g @ 40 cm)
6	Totally protected against dust ingress	Protected against high pressure water jets from any direction. Limited ingress permitted	
7		Protected against immersion between 15 cm and 1M	Protected against 6.0 joule impact (1.5 Kg @ 40 cm)
8		Protected against long periods of immersion under pressure	
9			Protected against 20 joule impact (5 Kg @ 40 cm)

Housings and their use

There are several tasks you will have to carry out which are common, whatever housing you opt to use for your electronic device.

The main tasks are:

- **MOUNTING** the printed circuit board

- **FITTING** controls and switches

- **FITTING** connections between the printed circuit board and the outside world

- **MOUNTING** other internal components besides the printed circuit board

- **WIRING** of all internal parts

- **LABELLING** the front and rear panels of the housing.

Mounting printed circuit boards

One of the primary purposes of a housing is to hold its internal electronic circuit's printed circuit board. It's not just a case of throwing the printed circuit board in, and closing the lid though. A printed circuit board should never be left loose inside any housing. Instead it should be properly and securely fixed, using one or more of a number of methods.

Often, the method you might choose to mount your printed circuit board inside your housing will depend on the type of housing you have opted to use *and* the size and shape of the printed circuit board. If you have a small printed circuit board and a plastic housing, it's perfectly possible to mount the printed circuit board to the inside of the housing using one or two pieces of double-sided adhesive sponge fixing pads, as shown in Figure 6.4.

FIGURE 6.4   Using double-sided adhesive sponge fixers to mount a printed circuit board inside a plastic housing: (left) applying the sponge fixing pads to the bottom of the printed circuit board; (right) applying the printed circuit board with sponge fixing pads to the inside of the plastic housing.

This method should not be used to mount a printed circuit board inside a metal housing though — for the simple reason that sponge fixing pads are thin, and some of the soldered connections on the bottom of the printed circuit board may touch the metal housing base (as shown in Figure 6.5), causing short circuits and possibly damage.

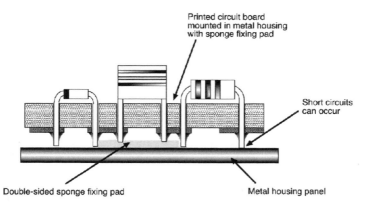

FIGURE 6.5   A printed circuit board should not be mounted inside a metal housing using sponge fixing pads, or short circuits may result!

Instead, when using a metal housing (or, indeed, if a larger printed circuit board is to be mounted even in a plastic housing) something more substantial must be used.

One of the best methods of mounting a printed circuit board in a housing is to use spacers. These are available in several formats. Some spacers use small bolts to hold them in place to the inside of a housing, as shown in Figure 6.6. Once the spacers are bolted to the housing, the printed circuit board is then bolted to the spacers. Obviously, the housing and the printed circuit board need to be drilled to accept the bolts prior to use.

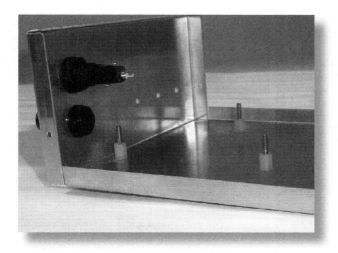

FIGURE 6.6   Bolted spacers, showing the spacers and the bolts.

Take Note

If you intend to use spacers to mount a printed circuit board to a housing, you must take into account where the spacers will be positioned on the printed circuit board. Component positions and the copper foil track may have to be adjusted to provide room – not just for the spacers themselves, but to allow access with tools to allow you to tighten the spacer bolts. You need to plan ahead, and decide which printed circuit board mounting method or methods you intend to use, as all the calculations regarding component positions and copper foil track layout need to be done in the very early stages of printed circuit board design – not at the last minute!

Other spacers are preformed, sprung-loaded plastic clips, which push into holes in the housing and the printed circuit board.

There are several variations of printed circuit board guides available. In use, these are fastened to the inside of the housing, and the printed printed circuit board is slid into position along them.

A particular variation of housing — extruded aluminium types — features in-built slots in internal surfaces. Conveniently, these are printed circuit board-sized, so provide an extremely easy method of mounting.

Take Note

In the printed circuit board design stages, make sure no components are positioned close to the edges of the printed circuit board if you plan to use guides of any sort. Further, if you plan to use an extruded aluminium housing, not only must there be no components fitted on the guide edge areas of the printed circuit board, but there must be no copper foil track in those areas.

If your printed circuit board has controls and connections (such as certain printed circuit board mounted potentiometers, switches and connectors) then you can use the controls and connectors to mount the printed circuit board inside the housing, with careful design. This is particularly so for small printed circuit boards.

Mounting the printed circuit board this way depends on using printed circuit board mounted controls and connectors. For example, Figure 6.7 shows a printed circuit board that has an on-board potentiometer and on-board connectors. These effectively hold the printed circuit board in place, as they are held to the housing sides.

Note though, that the actual printed circuit design is highly dependent on the actual controls or connectors. In other words, the housing you intend to use, the exact size and shape of the printed circuit board, *and* the actual controls or connectors all need to be taken fully into account during the very earliest printed circuit board design stages. Literally *everything* is dependent on each other. This method allows a highly integrated, a very professional looking, and often a much reduced size of device, so is often worth the extra time you might need to invest in the early design stages.

FIGURE 6.7   Using a printed circuit board mounted potentiometer and connectors to mount a printed circuit board inside its housing. This is the Guitar Headphone Amplifier project, in Chapter 8.

A common method of mounting small printed circuit boards inside housings is to make use of a particular type of connector known as an *edge connector*. Edge connectors are multi-pinned devices, comprising a row of contacts into which the edge of a printed circuit board fits tightly. A row of copper foil tracks on the edge of the printed circuit board mates with the edge connector contacts, as shown in Figure 6.8. The edge connector is bolted to the housing or to another — usually larger — printed circuit board. Thus, connections can be made to and from the printed circuit board, but just as importantly, the printed circuit board is held in position. A small printed circuit board mounted this way may need no other support.

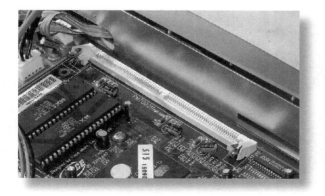

FIGURE 6.8   Using an edge connector to hold a printed circuit board in position. This particular connector (in white) is a RAM-card holder, common to computer and associated equipment.

Fitting controls and switches

While on the subject of controls and switches, we should include
details on how they should be mounted to a housing. The methods of
doing this are similar, whether they are special printed circuit board
mounted controls and connectors, or ordinary, more usual, controls and
connectors. The methods are also similar, whatever type of housing is
used — metal or plastic, or even wood.

Controls are required in most electronic circuits. They can range from
volume and tone controls in an audio amplifier, through time settings
for a digital clock, simple on/off switches, or dials to tune in radio
stations. The important point is that controls are used to provide an
interface between an electronic circuit (in printed circuit board form,
inside a housing) and the user of the electronic circuit (in human form
— normally — outside the housing).

Potentiometers

Most potentiometers have a threaded stem, that allows them to be
mounted to a housing using a supplied nut and washer. Figure 6.9 shows
a typical potentiometer, complete with washer and nut. A hole — the
diameter of the threaded stem — is required in the housing to insert the
stem. Also visible in Figure 6.9 is a location stub that extends beyond the
mounting point on the threaded stem. The purpose of the location stub
is to fix the potentiometer in position and prevent its rotation — as might
happen if the nut becomes loose, over time, say. The problem with the
location stub is that a second small hole (the diameter of the location
stub) is required in the housing alongside the hole for the threaded stem
— the correct distance from the centre of the threaded stem's hole.

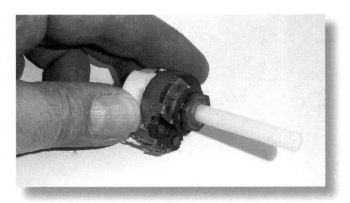

FIGURE 6.9   A potentiometer, showing its threaded stem, washer and nut, and the location stub.

Now, I should point out here that not all potentiometers have this location stub, and in which case the second hole is not required. And, indeed, I have been guilty myself of removing the location stub from a potentiometer in cases where I felt the nut wasn't likely to become loose, so didn't need the second hole.

Notwithstanding, the job to mount a potentiometer with such a stub is simple enough.

Mounting a potentiometer

- **MARK** the centre of the hole where the potentiometer's threaded stem is to be, and mark the centre of the location stub's hole, as shown in Figure 6.10. The distance between centres depends on the particular potentiometer, as does the angle at which you need to position the two holes relative to horizontal along the line of the housings panel.

FIGURE 6.10   Marking the position of the two holes required for the potentiometer of Figure 6.9.

- **MARK** both holes using a centre punch

- **DRILL** the holes using a pilot drill (no more than 2.5 mm). Position the housing over a block of wood, so that the drill bit doesn't grab the metal and distort it as the drill bit's cutting edge exits the hole.

CONTINUED...

...CONTINUED

- **DRILL** both holes using a drill the correct size for the location stub

- **DRILL** the threaded stem's hole using the correct sized drill

- **POSITION** the potentiometer in place

- **PLACE** the washer, followed by the nut over the threaded stem. Tighten the nut to a finger tightness

- **TIGHTEN** the nut, using pliers.

Take Note

When you tighten a potentiometer nut (indeed, any nut which is located on a housing's panel like this), you should take extra care to make sure the pliers don't slip off the nut and scratch the panel. Figure 6.11 shows how to tighten the nut. Note how the washer underneath the nut protects the panel from the pliers.

FIGURE 6.11   Tighten the potentiometer nut using pliers. Push the pliers against the washer under the nut to help prevent the pliers from slipping off, thereby protecting the housing panel.

Cutting a potentiometer stem to fit a control knob

- **PUSH** the control knob on the stem till it seats fully against the top of the stem

- **MARK** the stem directly in line with the bottom of the control knob (Figure 6.12)

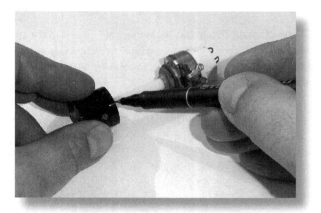

FIGURE 6.12   Push the knob on the potentiometer stem, and mark the stem.

- **REMOVE** the control knob, and measure the distance from the end of the stem to the mark (Figure 6.13)

CONTINUED...

...CONTINUED

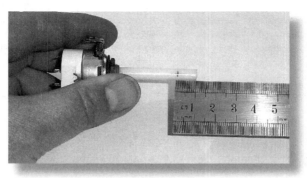

FIGURE 6.13   Measure from the stem end to the mark.

- **NOW** make a second mark on the stem the same distance from the top of the potentiometer's threaded stem. Actually, some knobs have an overhang specifically intended to cover the threaded stem, so this second mark can be a millimetre or two closer to the stem end if your knob features such an overhang

- **CUT** the stem on this mark. To do this, hold the end of the stem in a vice, with the stem horizontal. Hold the potentiometer body in one hand, while cutting the stem with a junior hacksaw in the other hand, as shown in Figure 6.14

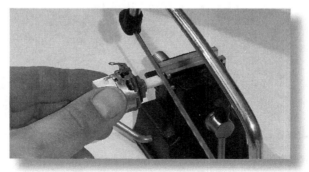

FIGURE 6.14   Cutting the stem of a potentiometer.

- **DEBURR** the cut end of the stem with a fine file. Your potentiometer is now ready to be fitted with its knob.

> **Take Note**
>
> Potentiometers are actually quite fragile. Do not hold the potentiometer body in a vice to cut off excess stem. Make sure you put the stem end in the vice!

Knobs

Many, many types of knobs are available, in all sorts of sizes and styles. Figure 6.15 shows a selection. Most are fitted to a potentiometer either with a push-fit, or (usually) with a small grub-screw through the side, which clamps onto the potentiometer stem.

Figure **6.15**   A selection of common control knobs.

Switches

Several variations exist in switches, with slide, toggle, push, and rotary being the most common (see Chapter 3). Within these variations, however, there are also variations in body shapes and sizes. As a result, it's not possible to include details of how to mount all switches here. Instead, I'm going to generalize, and show how the main variations should be mounted. These main variations are determined by body shape, not internal mechanics. Main body shapes are circular and rectangular, which narrows the field down somewhat.

Circular switches are rather like potentiometers in their fitting, and most of the details in the potentiometer section already covered are relevant here. Some circular switch types even have a location stub, usually in the form of a shaped washer, which requires a second hole (just as with potentiometers) to correctly mount the switch. Figure 6.16 shows a switch which mounts using this method. The method of mounting closely follows that of mounting potentiometers, earlier. Both holes should be marked out and drilled, then the switch is fitted using the location tab washer and a nut which fastens over the switch's threaded stem. As when tightening a potentiometer's mounting nut, take great care when tightening a switch's mounting nut, in case the pliers slip off and scratch the housing's panel.

FIGURE **6.16** A miniature circular toggle switch, complete with location washer having a location tab.

Some circular switches are provided with two mounting nuts (you can see one on the switch shown in Figure 6.16). These have the purpose of allowing the switch to be adjusted in terms of position, sticking far out from the housing's panel, or closer to the panel. A benefit of this second nut is that the outside nut needn't be tightened with pliers (thus avoiding any risk of damage to the panel outside), and all tightening is done to the inside nut, using pliers.

Rectangular switches are rather harder to mount, simply because rectangular holes are very much more difficult to err... drill. Of course, in general, it's a question of drilling a hole in the panel that is no larger than (but only slightly smaller than) the switch itself, then carefully filing out the hole till it is big enough for the switch to fit. The process is actually easy enough, just somewhat laborious, and follows now.

Fitting a rectangular switch

- **MARK** out the rectangular hole to be made

- **MARK** the centre, using a centre punch

- **DRILL** out a hole (using a pilot drill, and subsequent larger drills), just slightly smaller than the required finished hole

- **FILE** the hole, using fine files to the exact size required.

Hint

When you file a hole in your housing's panel, you'll need to clamp the housing in a vice. To protect the housing from the vice jaws, you should get two small blocks of wood and clamp the housing between the blocks. Make sure the blocks are positioned close to the edge of the hole you are filing – this prevents the panel from flexing too much and being damaged while you file. This is particularly important with thin aluminium folded-metal chassis-type housings. Also, file from the outside in – this means that any burring you naturally get while filing forms on the inside of the housing, so is not visible from the outside when finished.

Some rectangular switches have particular hole requirements. For example, slide switches (other than needing a rectangular hole for the switch tab itself to protrude through) also require holes to suit screws which hold the switch body in place. There's nothing for it but to carefully mark out the holes to drill and file, and get on with it!

▷ **FYI** ◁

Holes for some common rectangular switch shapes (and some circular ones, for that matter) can be cut out using hole punches, purpose-designed for the task. A hole punch comprises three main parts — a cutting tool, a socket receptacle, and a bolt arrangement. Using them is easy: you drill a hole the size of the bolt diameter, then fit the punch together, with the cutting tool on the front of the housing panel, the socket on the back, and the bolt going through the panel. As the bolt is tightened, the cutting tool punches out the hole in the panel, leaving a very clean hole. Such hole punches are quite expensive, but possibly worth the cost if you intend to make many identical holes.

> **Hint:**
>
> There are so many switch variations that it's usually possible to choose a switch type that is easy to fit. In other words, unless you want to use a rectangular switch for a particular purpose (although personally, I can't think of any!), my advice is to use circular switches always. Remember that every operation you perform to your housing's panel is just another opportunity to scratch the panel.

Fitting connections

Many electronic circuits require some form of electrical signal as an input, or an output. An audio amplifer, say, has outputs to loudpseakers, or a multimeter has inputs from its probes. The connectors used for such connections varely widely, depending on what sort of signals are being connected. Figure 6.17 shows a selection.

FiGURE 6.17   A variety of connectors used with electronic circuits and their housings.

The mounting of connectors to housings follows the same principles as that of mounting controls such as potentiometers and switches, with only a few minor additions or variations.

Often, circular connectors have a locating key of moulded plastic, which must mate with a notch cut in the panel as the connector is fitted to the housing. The purpose of the key and notch arrangement is simply to prevent the connector rotating. Figure 6.18 shows a connector with this sort of rotation prevention device. It all sounds good, but it does make fitting such connectors a little more tricky. After drilling a hole in the housing's panel to fit the connector, you then have to file out a notch in the panel to fit the key. A set of jewellers' files is best to do that.

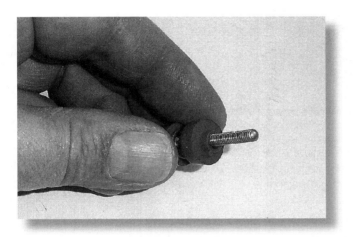

FIGURE 6.18   A connector with a location key – a corresponding notch in the housing panel must be filed (using jewellers' files).

Connectors need to be connected internally — usually to the printed circuit board, with a length of connecting lead. Some connectors may be multi-pole (ie, have two or more individual connections inside the one connector) while others will be single-pole (ie, only one individual connection). The type of connector defines what sort of connecting lead you need to use.

Often multi-pole connectors are audio in nature (microphone, earphones and so on) and they will usually require a special type of connecting lead known as screened cable. Screened cable is formed with a central insulated lead (some screened cable has more than one central insulated lead), wrapped with uninsulated wire, and the whole construction is insulated. Figure 6.19 shows some screened cable, in which you can see the central insulated lead and the uninsulated wire. The uninsulated wire around the central lead is intended to be earthed, so offering a measure of protection (or *screening* — hence the name) for the relatively small signals carried by the cable against electro-magnetic interference.

FIGURE 6.19   Screened cable, showing a central insulated lead, and the wire screening, all encased in insulation.

A variant of screened cable, for use with high frequency signals, is coaxial cable. Like screened cable, coaxial cable has a central signal lead, with an external screening. However, the screening is usually in a braided form.

Usually, connecting leads between the connectors and the printed circuit board are made by soldering the lead at each end. However, some connectors (notably power connectors) require some other joining method. The connector of Figure 6.18, for example, features a bolt forming the internal connection. Such a connector requires that solder tags be soldered to the ends of the connecting leads (Figure 6.20 shows a connecting lead with a solder tag connected to it). The solder tag can then be bolted to the connector.

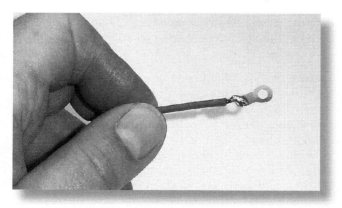

FIGURE 6.20   A solder tag, soldered to connecting lead, for use with bolted connectors.

Other power connectors — usually mains connectors, although the connection method is popular in car electrical circuits too — feature a spade male lug, as shown in Figure 6.21. A mating spade female connector must be crimped to the connecting lead, using a crimping tool. The result is a solid and secure connection.

FIGURE **6.21** *Connector featuring spade lugs.*

Some connectors mount directly on to the printed circuit board, which negates the need for leads between the connector mounted on the front panel and the printed circuit board. As a benefit such connectors can be used, as shown on page 148, to aid when mounting the printed circuit board itself in the housing.

Mounting other internal components

A few other internal components that need to be mounted on a housing exist. The most common group is that of indicators — LEDs, bulbs and so on. Generally speaking, they have the same mounting conditions that connectors and switches have, with the exception of LEDs.

Several varieties of LEDs exist. Simplest to mount are the basic, round common LEDs, as shown in Figure 6.22. These *can* be mounted just by drilling a hole in the housing panel the exact diameter of the LED and pushing the LED in.

The problem with this method is that it's just as easy to push the LED back out, from the outside of the panel. The solution is a simple clip, shown in situ on a LED in Figure 6.23. In use, it's a simple matter of drilling a hole in the housing panel the correct size for the clip, inserting the clip into the hole from the panel front, then pushing the LED into the back of the clip, until it seats correctly with a click. Clips come in the same range of sizes that LEDs do — it's just a matter of making sure you have a clip the same size as the LED.

FIGURE 6.22 Simple mounting of a LED to a panel.

FIGURE 6.23 A LED, complete with panel clip.

Fuse holders

Mains-powered circuits usually require a fuse for safety purposes, and the most common type of fuse holder is chassis-mounted. These are fitted to a housing panel, so that the fuse is accessible from the outside of the housing, while all mains connections are made on the inside (see Figure 6.24). There are no particular considerations in the actual mounting requirements, but there may be wiring precautions (to make sure live terminals cannot be touched when removing or installing a fuse).

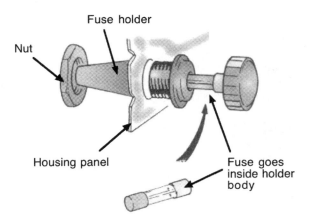

Fuse holder

Nut

Housing panel

Fuse goes
inside holder
body

FIGURE 6.24   Mounting a fuse holder into a panel.

Heatsinks

All electronic components generate heat as electric current flows through
them. Generally this is not a problem, as they naturally dissipate the heat
so that the temperature doesn't get too high to damage them. However,
large current flow through certain components such as certain transistors,
generates heat to the point that such a component *could* be damaged if
the heat was left to dissipate naturally, and some method of removing
heat is required to maintain a component core temperature lower than
that which would cause the damage.

The method of removing heat — known as *heatsinking* — usually relies
on mechanically connecting the component to a large mass of metal. As
metal is a heat conductor, heat from the component core is thus shunted
away, preventing damage.

Inside an electronic device's housing, components which may heat up
to the point of damage can be protected with readily available push-,
clip-, or bolt-on heatsinks; a selection is shown in Figure 6.25. Typically,
the larger metal surface of the heatsinks allows a degree of cooling just
because of increased natural convection. To increase the surface area
and air flow around heatsinks, they are often finned.

Most heatsinks of this type are simply fitted to the component in question
and then forgotten about, so they provide quite an easy way of providing
component cooling. They do, however, only have a limited ability to aid
component cooling.

F𝘪ɢᴜʀᴇ 6.25   A selection of small heatsinks.

Where components are likely to get hotter than such heatsinks can cope with, a common trick is to mount the components on a panel of the housing (assuming the housing is metal, of course).

All these heatsinking methods rely on the fact that the medium directly in contact with the hot component will dissipate heat sufficiently to maintain an acceptably low temperature. When deciding which heatsink is required for which component, there are five main variables which need to be considered.

Five variables of heatsinks

• **TEMPERATURE of the component's surface.** What is the maximum temperature the component can reach when operating under circuit conditions?

• **TEMPERATURE of the surroundings.** What is the maximum ambient temperature?

• **THERMAL resistance of the junction between the component and the heatsink medium.** How easily does heat flow from one to the other?

• **ABILITY of the heatsink to dissipate heat**

• **ELECTRICAL conductance.** Components are metal, heatsinks are metal, metal conducts electricity — so is there likelihood of short circuits?

If you are designing an electronic circuit from scratch, then you need to know or at least be able to calculate all the above variables to ensure that a heatsink will do its job properly. If you are simply building a circuit from a magazine or book such as this one, then you don't need to do any calculatons — simply use the heatsink that is recommended.

Whether you design a circuit yourself, or follow one, there is a particular method when using a heatsink to make sure the best performance is obtained.

The problem is that the surface of a component, and the surface of a heatsink are not always perfectly flat. Small air pockets always exist which restrict heat flow from the component to the heatsink — known as the *thermal conductivity*.

Thermal conductivity can be improved by coating both surfaces with *thermal compound* (sometimes known as *thermal joint compound*, or *heatsink compound*, or *thermal grease*). This is a silicone oil based grease, impregnated with metal (usually zinc or silver) particles, which fills in air pockets and promotes heat flow. It's usually supplied in tubes or syringes (see Figure 6.26) for easy application.

Using it it quite easy: apply a small amount to your finger, then rub your finger and thumb together to thin the compound out, then apply some to each surface.

Finally, mate the surfaces.

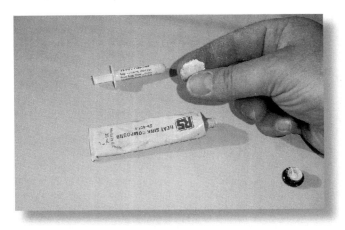

FIGURE 6.26   Thermal compound – supplied in various ways. Put a small amount on your finger and rub between finger and thumb before applying to a component or heatsink surfaces.

Take Note

It is important not to have too thick a layer of thermal compound between component and heatsink! Too much compound restricts heat flow just as much as too little compound.

▷ **FYI** ◁

Even mounting components on a metal housing's panel may not provide sufficient cooling for some applications. Better cooling methods need then to be adopted — the most common of which is the fan (seen most often in large computers) which helps to ensure that air flows round hot components. Other heatsinking methods exist, including heat exchange, refrigeration and water-cooled systems, but these are generally beyond the scope of the beginner to electronics construction.

To make sure no electrical connection occurs between a component or its heatsink, use an insulating washer and plastic mounting tab or bush. Traditionally, insulating washers are commonly made of the insulator mica, but increasingly popular are silcon rubber washers. Both are shown in Figure 6.27 along with an insulating mounting tab. Note that the size and shape of the insulating washer depends totally on the size and shape of component it is for. The washers shown in Figure 6.27 are meant to fit a TO220-type transistor shape component. Insulating washers — obviously — fit between the component and the heatsink, so introduce another two surfaces into the equation, so when applying thermal compound prior to fitting a heatsink, put thermal compound on all four surfaces.

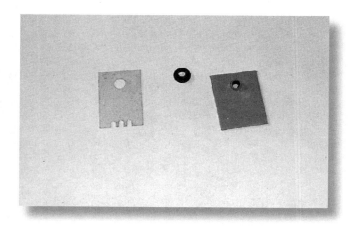

FIGURE 6.27   Insulating washers: left is a mica washer, right is a silcon rubber washer. In the middle is an insulating mounting tab.

The component needs to be fastened mechanically to the heatsink in some way. When fastening a component to a housing panel the usual method is with a small nut and bolt. The plastic insulating mounting tab ensures that the nut and bolt used do not cause a short circuit between the component and the housing panel. The way the arrangement should be fitted together is shown in Figure 6.28.

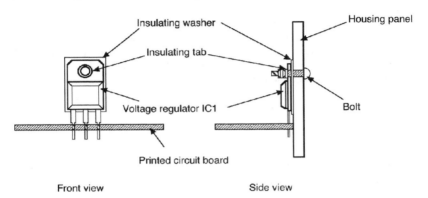

Front view   Side view

Figure 6.28   Bolting a component – in this case a voltage regulator integrated circuit – complete with insulating washer and mounting bush, to a heatsink.

Figure 6.29 shows an actual application of heatsinking components, in the Bench Power Supply project of Chapter 8. Here you can see the three voltage regulators that are heatsinked to the rear panel of the project housing.

Figure 6.29   Showing three components, heatsinked to a metal housing rear panel.

> ### Take Note
>
> Electrical conductance does not need usually to be taken into consideration where a heatsink is a simple push-, clip-, or bolt-on type, that is fastened directly to an individual component on a printed circuit board. After all, as long as the heatsink touches nothing else metallic, there can be no danger of short circuits.
>
> However, electrical conductance for a component mounted onto a metal housing's back panel for heatsinking purposes is of prime importance — several components of circuit parts (the printed circuit board, connectors, controls, power supply parts and so on) may also be mounted on the housing. Here, problems of possible short circuits need to be taken seriously. Some form of insulating washer should be used to prevent electrical connection between the component and the housing.

Wiring of all parts

In some instances, once a printed circuit board is mounted in its housing, the electronic device is complete, and it only requires the housing lid to be fastened on for final completion. In most instances though, there will remain some wiring of internal parts together.

If you are building an electronic device from a magazine or book design any internal wiring will usually be shown as a wiring diagram, although this is normally topographic — ie, like a railway map, it shows theoretical connections but not the actual routes. More often than not, how the wires themselves are connected between parts is left to you, the constructor to determine. With this in mind, there are some basic guidelines, but the rest is up to you.

Generally, use multi-stranded insulated wire for connections.

Multi-stranded wire types and uses

- **FOR low voltage, low power connections a suitable choice is known as 7/0.2 equipment wire (which means it has seven strands of wire, each one being 0.2 mm in diameter).** This is flexible and easy to route inside housings.

CONTINUED...

...CONTINUED

- **AN alternative to individual wires is ribbon cable (ie, several strands of insulated 7/0.2 wires, moulded together side-by-side).** Ribbon cable has an advantage over single equipment wire in that where multiple connection are made between internal parts, connections are grouped. Ribbon cable is available with each strand a different colour, which makes checking connections an easy job, too. Ribbon cable is available in 10-, 20-, and 30-way strips, but strips can be easily separated just by pulling, into the number of connected strands you want for each connection

- **MAINS or higher power connections must be made with thicker multi-stranded wire (eg, 16/0.2, 24/0.2, or 32/0.2).** Often the magazine or book will specify the exact type to use

As construction of the circuit and wiring goes on there may be instances where several lengths of wires follow the same route. As a matter of neatness, and to prevent wire movement (which could cause soldered joints to come apart) it's a good idea to tie the lengths together. Plastic cable ties are ideal for this job. An alternative is the cable clip, which comes with a double-sided adhesive pad to stick the clip onto a housing panel. Use ties or clips at least every 100 mm or so along panels, and at corners where the grouped wires go round.

▷ **FYI** ◁

In the 'good old days' tying of cables was done with special continuous cable tie cord, and was more a work of art than a technique. Fortunately, 'good old plastic' cable ties do an equally good job these days.

Where wires are grouped and tied or clipped together into cables, it can become tricky working out which wire is which. If you have a selection of coloured wires this can help, as can using coloured ribbon cable. But where you have more than just a few of these, you still might find it awkward. If this is the case, use cable markers. These are small plastic rings, placed over each end of wires before soldering. They can be coloured, and also numbered.

Mains wiring

The wiring of electronic circuits which use mains power bears particular mention. Mains power is, of course, dangerous — potentially lethal — and precautions must be taken when you intend to build a circuit which uses it.

General guidelines for mains wiring

- **IN general, use a metal housing for any electronic circuit you construct that has a mains power supply.** Use three-core mains cable (ie, with live, neutral, and earth wires), and make sure the cable enters the housing using a good quality cable entry clamp, properly tightened so that connections are not strained. A selection of cable entry clamps is shown in Figure 6.30. You must ensure the clamp is the correct size for the cable used.

FIGURE 6.30   Cable entry clamps for use when mains cable enters a housing.

- **AN alternative to a cable entry clamp is a good quality plug and socket connector — such as that shown in Figure 6.21.**

- **USE insulating sleeving over all soldered connections, to ensure that no mains electrical parts are exposed.** Push the sleeving onto a wire prior to soldering it to a connection. After soldering push the sleeving back along the wire, and over the soldered joint, as in Figure 6.31.

CONTINUED...

...CONTINUED

FIGURE 6.31   Using insulating sleeving over a mains circuit soldered joint. Here the joint has been made, and the sleeving is being pushed into position to cover the joint.

- **ALL soldered connections must be made good mechanically, prior to soldering.** Thus, if the soldered conection comes loose, the wire is still held in place.

- **MAINS live and neutral wires should go straight to an on/off switch.** Use a double-pole on-off switch for this purpose, and connect the live connection to one pole, and the neutral connection to the other. Such a switch ensures that the following circuitry is completely disconnected from mains when the switch is switched off.

- **TO protect the electronics circuit (and you, for that matter) from damage in case of an internal fault, use a panel-mounted fuse holder, with a fuse rating no greater than that specified in the magazine or book the electronic circuit comes from.** Mains live is taken directly from the other contact of the live pole of the on/off switch to the central connection of the fuse holder (thus, if the fuse is removed when the device is connected to mains, no contact can be made to a live part).

CONTINUED...

...CONTINUED

- **THE earth wire in the mains cable must go directly to a bolted connection on the housing chassis.** Use a solder tag on the earth wire, and shakeproof washers on the bolt, to ensure the bolted connection is reliable over the long term.

- **CHECK,** double-check, and triple-check all connections and the circuit prior to connecting to mains power.

Figure 6.32 shows a general wiring diagram for mains powered electronic circuits which follows these guidelines.

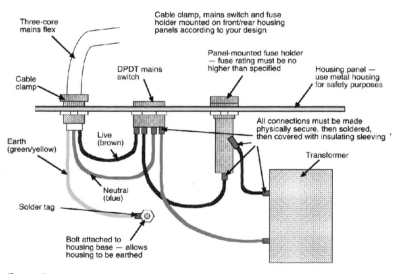

Three-core mains flex

Cable clamp, mains switch and fuse holder mounted on front/rear housing panels according to your design

Panel-mounted fuse holder — fuse rating must be no higher than specified

DPDT mains switch

Housing panel — use metal housing for safety purposes

Cable clamp

All connections must be made physically secure, then soldered, then covered with insulating sleeving

Earth (green/yellow)

Live (brown)

Transformer

Neutral (blue)

Solder tag

Bolt attached to housing base — allows housing to be earthed

FIGURE 6.32   A general wiring diagram for mains powered circuits.

Take Note

If a single-pole switch is used, even if it is in the live circuit on construction, at some stage somebody may rewire the mains plug incorrectly, at which point the switch is then in the neutral circuit and the remaining circuit will be live even if the switch is turned off! A double-pole switch prevents this situation.

Front and rear panel labelling

When an electronic circuit is mounted in a housing, and all connectors and controls are fitted, then the device is ready for use. Well, not quite!

The main interface between you and the electronic circuit you've just made is the housing's front panel and all its controls and connections. With maybe just a couple of these, you probably remember what each is, or does. But if there are more than just a couple, it becomes increasingly difficult to remember. And if anyone else needs to use the device, how will they even know at all?

The answer is that your front panel (and possibly the rear panel if there are controls and connections round the back) should be labelled.

Labelling is all part of the process which is often called *finishing off*. The point is though, that far from being a process you undertake at the finish of building a circuit, you need to be considering the front and rear panel and its layout right at the very earliest design stages.

For a start, labelling can't possibly be the last thing you do — to put labels on, all controls and connections need to be removed so that you can work on a clean, flat panel. So, when you think about it, Labelling is not actually a finishing off process, it has to be done quite a bit earlier on in the whole build.

Panel labelling

Once you know the layout of the panels, and have drilled holes for all control and connectors you can start labelling. There are several methods for this. The simplest would be merely writing the labels on with a permanent ink marker pen, but of course that would not make a particularly good-looking end result.

A much better method however is to use rub-down transfer letters. These are available in hundreds of typefaces (more commonly called fonts and sizes), and it is quite simple to create a good effect with a bit of practice.

Transfers such as these are available in sheets, and any good stationers will stock them or order them for you. Ask to see their catalogues to see the range of typefaces available.

Key points when labelling with rub-down transfers:

- **USE** guidelines to ensure labels are level. Rule faint lines in pencil, or stick lengths of tape along the length of the panel

- **TAKE** care when centring labels around controls or connections — if they go too far to one side or the other they will look odd

- **GROUP** controls and connectors where you can. Perhaps ring them using some rub-down transfer lines, so that they appear cohesive in one part of the panel

- **USE** a typeface of only a few millimetres high for most labels. This equates to a publishing measurement (in points), of around 12 to 16 points

- **USE** a plain typeface for most labels. In publishing-speak, this could be a sans serif font

- **USE** a more ornate typeface if you want for the main descriptive label — eg, Audio Amplifier.

Using a computer

If you have access to a computer, you can use it to aid when labelling a housing. First you can design the panel layout on screen, simply to give yourself a good idea of what looks good — even if you opt to use rub-down transfer lettering to actually do the labelling in the end. All you need is a fairly basic illustration application of some description on your computer. There are several such applications available, and even a fairly basic image manipulation application (such as you probably have bundled with your printer, for working with photographs, say) can often produce good results. All it needs to be able to do is to use the range of fonts and sizes your computer already has installed, and perhaps draw a few lines. Of course, a proper illustration application is best.

Designing a good-looking label is not rocket science — the key points above can help. The advantage of designing the panel on a computer screen, though, is that you can make adjustments until you get it just right. Figure 6.33 shows the sort of label you can hope to create using a computer. Note the outside marks, which act as registration marks to make sure the label will fit the panel exactly.

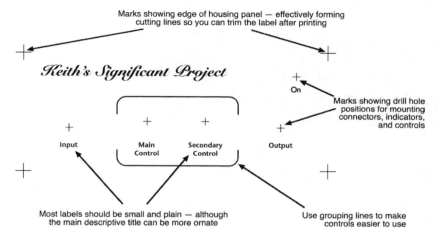

FIGURE 6.33   A computer-generated front panel.

Better than this though, after you have designed the panel label on screen, you can print it out using a printer, then attach the whole label to the housing panel in one step. You don't even need a particularly expensive printer to do this — most modern inkjet printers produced over the last few years would work. As you are printing the label, you are not restricted to just monochrome, and judicial use of colours can be used to make the housing more attractive.

Simplest way here is to print the design onto photographic paper (glossy or matt, whichever you prefer). The paper can be trimmed to size, then stuck to the housing panel using adhesive (a spray adhesive is ideal for this task).

Alternatively, if your printer is capable, you can print the design onto acetate transparency. Print the design in mirror-image format on the back of the acetate sheet, then trim the acetate to fit. Stick the acetate onto paper (white or coloured, for the desired effect), then stick the whole thing onto the housing panel.

One of the best methods of labelling using a computer is with the Quick-Laser system, shown in Figure 6.34. In this process, special coloured polyester sheets with optional, die-cut and peel-off sections (similar in principle to address labels for envelopes) are used together with a laser printer to produce very high-quality self-adhesive labels.

FIGURE 6.34   The Quick-Laser system uses coloured polyester sheets to produce labels from a computer and laser printer (Mega Electronics UK).

House!

Finally, after you've labelled the housing, you can refit the controls and connectors, and your electronic device is complete. Congratulate yourself!

7

Fault-finding

At some time or another, anyone and everyone who has more than a passing interest in electronics has built a circuit up, perhaps spending a great deal of time not to mention money in doing so, only to find that it doesn't work. Even the professionals encounter this, so what chance have people like you or I got in producing a working and reliable circuit?

Well, fortunately, we have a remarkably good chance of doing so, particularly if we follow a fairly logical and straightforward approach to *building* electronic circuits in the first place. Fortunately also, in the vast majority of cases the reasons why circuits don't work is that there is some form of easily identifiable mechanical fault.

So, by a combined strategy of building electronics circuits properly in the first place (so that they *do* work first time usually), and being able to identify and correct the cause or causes of them *not* working in the second place, then we all stand an extremely good chance of being able to produce working circuits.

That is what this chapter, indeed, what this whole *book*, is all about — producing a combined strategy of proper building techniques so that (hopefully) circuits *will* work first time, together with the means of identifying the problems and faults in those circuits that do not work first time.

> **Hint**
>
> Remember: fault-finding is not just about locating what you've done wrong after you think you've finished — it's about locating what you did wrong BEFORE you think you'd finished...

A strategy of building electronic circuits

With all this in mind, it's obvious that to prevent faults in a circuit which you will have to spend time fault-finding later, it's better to take care when building your circuit in the first place so that the faults simply don't occur. Remember the old saying: a stitch in time saves nine.

As such, I'm going to give you, first, a fairly straightforward list of things that you should do when building an electronic circuit which, if you *do* follow will help to largely negate the need for fault-finding afterwards. This list is based on a knowledge of what faults typically do occur in electronic circuits, and what's more, why those faults occur. At the end of the chapter, I'll then give you a method of fault-finding which will see you in good stead for the rest of your electronics constructional life.

A Strategy To Follow

KEEP CLEAN — every part of the electronic circuit. Printed circuit board track, component leads and so on can all be affected by dirt and grease to the extent that soldered joints are not properly made. Worse, dirt or grease in a joint can affect its performance over time — it might work initially but stop working after a short while. Often, just cleaning the parts is enough to ensure a good joint.

CLOSELY INSPECT — the printed circuit board. Your complete electronic device depends totally on the printed circuit board. A minor imperfection (such as a crack in the copper foil track), or a hairline short circuit of copper between tracks which are not intended to be connected will prevent a circuit from working.

DOUBLE-CHECK — component values. Check once when you choose a component that it is the correct value, then check it again after it's been inserted in the board immediately prior to soldering.

DOUBLE-CHECK — polarized components. Electrolytic capacitors, semiconductors and so on must be inserted into a circuit the right way round (this is called *correct polarization*). Check their polarity is correct once when you insert the component in the board, then once more immediately prior to soldering.

TRIPLE-CHECK — component positioning. As you select a component for insertion into a printed circuit board, check its value on the circuit diagram. Next, check that the position on the printed circuit board is correct according to the layout diagram. Finally, after insertion, recheck both value and position again immediately prior to soldering.

DO NOT — insert more than a handful of components in at a time. If you insert too many components in to a printed circuit board before soldering them in, they may slip partially out of the circuit board. This may create short circuits that affect the circuit's operation. Only insert a few components before soldering them in, then moving on to the next batch.

WHEN SOLDERING — make sure each and every joint is properly made. Check each joint visually immediately after making the joint. Do not make hundreds of joints one after another without checking. Read Chapter 5 thoroughly, and practise making soldered joints until you can make a good joint *every* time.

AFTER SOLDERING — but before testing. Check for shorts between joints, and check for dry joints.

BUILD STAGES — and test as you go along. Divide the complete circuit into complete functional units (most — if not all — circuits can be thought of this way) and build each stage up in turn. As you finish one stage you can test it, before moving on to build up and test the next stage. In this way, if you *do* have a fault, you have isolated it in the production stage, and so can repair it more easily than if the complete circuit is built.

POWER ON — as a last resort only. Inevitably, when you have finished your printed circuit board you want to connect power and see if it works. The problem is — if you do and there is a fault such as a short circuit, or if polarized components have been inserted the wrong way round, or if components of the wrong value have been inserted, or for any number of other reasons — you could easily cause irreperable damage to components. Take time out for a few minutes, then inspect the printed circuit board carefully — use a magnifying glass if you have one, to see if you can find any possible faults. Remember: once you've connected the printed circuit board to its power supply it might be too late — a simple fault becomes a big-time problem...

Types of faults

If I was to be pinned down on this issue, I would have to say that in 99.999999999999999 (OK, that's enough!) per cent of cases, the reason (or reasons) why a circuit does not work is due to a mechanical fault introduced by the person making the circuit. Very rarely — and I do mean *very* rarely — the reason why a circuit does not work properly first time is due to a faulty component. This is, quite simply, because electronic components are made to a pretty high standard, and are quality controlled so that you may rarely expect one to not work. Even more rare are those faults which occur due to a circuit (an incorrectly specified component, say, or an error in circuit design).

As a result, when a circuit you have built does not work, the first and most important thing to do is assume *you* are at fault — not the components. The second thing to do is look closely at what you have done — to find out where you have gone wrong.

With this in mind, you can now try to isolate the reason why the electronic circuit you've built doesn't work. Check things in the following order:

Finding faults? Walk this way:

- **SOLDERED** joint faults

- **PCB** faults

- **WIRING** faults

Once you've thoroughly checked, and are happy that your soldering, your pcb and your wiring is not at fault, then — and only then — you should begin to consider the remaining potential causes of faults:

- **CONNECTORS**

- **COMPONENTS.**

Soldered joint faults

Basically, there are two types of faults that can occur when soldering. The first is just a bad, or dry joint. The second is a solder bridge. Figure 7.1 compares a good joint with a bad joint and a solder bridge. Good joints tend to be concave and shiny in appearance. Bad joints tend to be blobby (to the point of being ball-shaped) and may be dull in appearance. Solder bridges form between copper tracks on the printed circuit board. In many cases of bad or bridged joints you can isolate each merely with a close visual inspection. It helps, though, if you have a magnifying glass and good lighting.

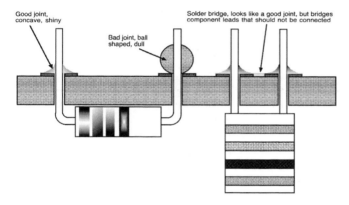

FIGURE 7.1   A good joint, a bad joint, and a solder bridge.

If you spot bad joints, they will need to be resoldered. Note that one of the main causes of a bad joint is that the joint was not hot enough before solder was applied, so when resoldering a bad joint, make sure it gets sufficiently hot first.

If you spot bridged joints, you need to reheat the joint and remove the excess solder. Use either desoldering braid or a desoldering pump for this task, but remember that once the solder has been removed you will need to reapply fresh solder to make the joint again.

PCB faults

Particularly if you make your own printed circuit boards — although even if you buy them ready-made — you should be aware that they can have faults. So, no matter how good your soldering or circuit building techniques are, they will never be able to produce a working electronic circuit. A good procedure would therefore be to check the printed circuit board, before you even start to solder components into it.

Look for three things:

- **1 Bridges between adjacent copper tracks.** Sometimes, if the etching process hasn't fully completed before the printed circuit board is removed from the etchant, or if small areas of etch-resist have not been developed fully, small bridges of copper remain between tracks. Sometimes these bridges can be extremely thin, but even a hairline bridge forms a short circuit. Thoroughly check all copper tracks of the printed circuit board — a magnifying glass and good lighting helps. If you spot what you think is a bridge, you can confirm it using a multimeter set to resistance measurement. Bear in mind that if you find one copper bridge, there is a good chance there will be more at other places on the printed circuit board

- **2 Broken tracks.** In the etching process, a small (even minute) break in the resist allows the etchant to remove the copper underneath. If the break in the resist cuts across a track, there will be a complete break in that track. The circuit built on that printed circuit board will not work correctly. Again, there is nothing for it but to first check the copper track visually. If you spot a broken track, clean the area to either side of the break with an abrasive scrubbing block or abrasive fibreglass pencil, then solder a joint which bridges the two sections of copper track

- **3 Incorrect or missing track.** The copper track on your printed circuit board is intended to represent and make all connections between components in the circuit diagram. If you have designed your own printed circuit board from a circuit diagram then made it yourself, there is the potential that somewhere in the copper foil track layout design process you have made an error. If the completed electronic circuit does not work, then as part of the fault-finding process you need to double-check your printed circuit board's copper foil track layout. Make sure that all connections on the circuit diagram have a corresponding connection on the copper foil track — if any connections are missing, then the circuit parts are effectively open circuit. If you find any connections missing, make them using a small length of 7/0.2 wire. Strip a very short amount of insulation from each end of the wire, then tin each end before soldering it into place to make the missing connection. Also make sure there are no *extra* connections (possibly forming what amounts to short circuits between circuit parts) that should not be in the copper foil track. If you find a connection present that should not be, cut through the copper track using a craft knife or similar to break it.

Wiring faults

When constructing any electronic device, mounting the printed circuit board in a housing, then connecting the printed circuit board to controls and connectors, wiring is always an issue and a potential area for faults to arise. If you suspect wiring to be the problem (or, at least, you have gone through the processes above and the circuit *still* does not work), check the wiring against the wiring diagram. Isolate each connection in turn between the printed circuit board and housing-mounted control and connectors, and make sure each end of it connects to its correct places. Look out for bad soldered joints at each end. Note that some connection terminals on connectors are relatively large metal terminals (at least in comparison with copper foil track connections), and soldering a connection to them requires significantly more heat than it does to printed circuit board joints. As a result, it's easy to make a bad joint to such terminals. Resolder any joints that look in any way suspect.

Connectors

After soldering errors, printed circuit board faults, and wiring errors, the next most important cause of faults in electronic devices is that due to connector faults. Most of these occur after time and use, but you still must consider connectors as a potential cause for faults if you have worked through all the processes covered already in this chapter.

It is often easy to check that a connector is doing its job by using a multimeter set to resistance measurement mode. If resistance is infinite between one end of the connector and the other, the connection is open circuit. If two or more terminals on one side of the connector have zero resistance between them, the connections are short circuited.

Components

Finally, once you have gone through all the processes above to check soldered joints, printed circuit board copper foil tracks, wiring, and connectors, you can begin to suspect that one or more of the components is faulty.

Bear in mind though, that it may even be that one of the faults you have already found as you work through the processes in this chapter has actually caused one or more of the components on the printed circuit board to be damaged. A short circuit due to a solder bridge or copper track bridge, say, might cause sufficient current to pass through a component not designed to cope with the current to be damaged. So, even though you might have found one fault (or more) in your soldering or the printed circuit board, or the wiring, others may have been created by this.

Down with fault-finding

It's not the intention of this book to cover all fault-finding processes when all the processes have been followed and we are finally down to component level. There are, though, many other books on this topic. Just remember that fault-finding at component level must be the very last resort.

Instead, my intention in this book is to help the beginner to the construction of electronic circuits and devices see that fault-finding is not a process distinct and separate from the construction processes. Fault-finding should be, in fact, completely integral with the constructional processes.

Indeed, it should be *so* integral that checking for faults and correcting them as they occur should be natural to the constructor. So integral that you shouldn't even *think* of fault-finding as a distinct and separate process at all.

With this in mind, shown opposite is a method of fault-finding which comprises a general list of things to do and bear in mind as you construct any electronic circuit or device. As you become more experienced in electronics construction, the list should begin to become second nature to you.

Picture this!

So far, we've only looked at the theory behind fault-finding, trying to isolate the areas where faults can arise, what sorts of faults there may be, and what causes them. For the rest of this chapter we can take a look at what these faults actually look like.

Fortunately, several mechanical printed circuit board faults can be fairly readily identified with a close visual inspection. However, for the beginner to electronics construction (and, indeed, for even the not-such-a-beginnner!) it's not always *easy* to identify them, simply because finding something when you don't know what it looks like is like looking for a needle in a haystack when you're blindfolded.

So, for the remainder of this chapter, some photographs of the most common printed circuit board soldering faults are given, in Figures 7.2 to 7.14.

A total method of fault-finding:

- **KNOW** the circuit, and how it works. Break it up into logical *black box* sections. Often, the magazine or book giving the circuit will do this for you, in a separate 'How it Works' section, but if not, try to do it yourself. The more you understand how a circuit works, the easier it becomes to construct and get working!

- **CHECK** the plain printed circuit board. Check it for any short circuits. Check it for any broken tracks. Check that it follows the circuit, with copper foil track for every connection within the circuit, and with no extra connections

- **BUILD** one section of the circuit at a time. Insert and solder only those components within that section, then check the section works fully before you start the next section. If one section does not work, find and correct the fault before you start the next section. This way, any faults should lie only in the last section you built

- **ONCE** the printed circuit board is complete, with all components 'stuffed', check that it works. If not, as long as you have been following this method, the fault should again lie in the last section you built

- **HOUSE** the printed circuit board, together with controls and connectors

- **CHECK** the complete and housed device.

Blistering or lifting — where copper track lifts from the board surface, usually due to excessive soldering heat.

FIGURE **7.2** Lifted pads on a printed circuit board.

Bridging or shorts — where solder joins two or more conductive parts which are not meant to be connected.

FIGURE **7.3** Solder bridging between connection pins of an integrated circuit.

FIGURE 7.4   A minute solder bridge between two component pads.

FIGURE 7.5   Solder bridging between adjacent printed circuit board tracks.

Components lifting — where components become displaced between stuffing and soldering. Faults like these don't necessarily cause the circuit to malfunction by themselves, but components that do lift may short to adjacent components.

Figure **7.6**   A capacitor that has lifted.

Figure **7.7**   A lifted resistor.

FIGURE 7.8   An integrated circuit that has lifted.

Cracked joint — if a printed circuit board is not adequately fixed in position in a housing, occasionally soldered joints that come under mechanical stress can crack, thereby creating a broken connection. The hairline cracks so caused are very easy to miss under normal visual inspection — a magnifying glass is usually required to see them.

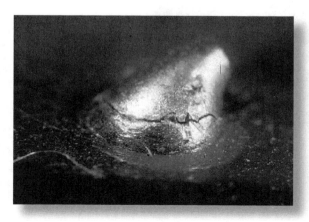

FIGURE 7.9   A cracked joint.

FIGURE 7.10   Another cracked joint.

Dewetting — where an initial bond is formed when soldering, followed by a withdrawal of solder from the joint, leaving irregular shaped mounds of solder separated by areas covered only with a thin solder film.

FIGURE 7.11   Dewetting between two connected component lands.

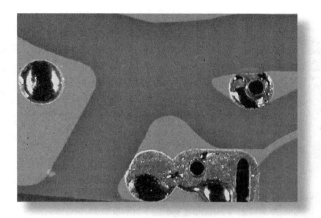

Figure 7.12   Dewetting of component lands.

Excessive solder — too much solder on a joint — this may also cause bridging with local lands.

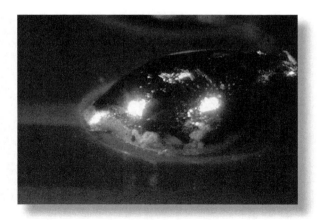

Figure 7.13   Excessive solder looks like a blob on the printed circuit board.

Solder voids — too little solder on a joint creates a potential weakness.

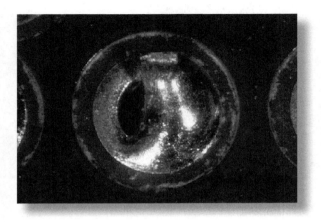

FIGURE 7.14   A solder void on a component land.

8

Projects

While this book so far has been quite theoretical in parts, I hope I haven't lost you all to sleep...

To wake you up again if you did succumb to the effects of boredom while I've waffled on, I've included some projects in this chapter that you can choose to construct if you want.

Even if you choose *not* to construct them, the projects are still highly useful, in that they illustrate many of the processes that I have covered throughout the book. They are graded to a limited extent — the first ones being easiest to build and the later ones being harder. That's not to say that if you want to construct one of the later circuits first you shouldn't be put off by its greater practical requirements — with the exception of the last project — the mains-powered Bench Power Supply. The principles, after all, of contructing any electronics circuit are the same whatever the complexity, and are covered in detail throughout this book. So, if you decide to build any of the projects, hopefully all the information you need to do so is somewhere in these pages for you to read.

Take Note

The last project — a mains-powered power supply — should not be attempted unless you are confident of your skills. Mains voltages are dangerous — potentially lethal — and extreme care must be exercised when constructing the project to prevent a situation arising where a user comes into contact with those voltages.

In short, if you are not confident of your skills, and do not take requisite care — do not construct this project!

The projects

I thought long and hard about these projects, trying to weigh up complexity, ease-of-construction, and usefulness, aiming to get a balance between all three. What I decided early on was to promote what I feel is the *best* way of constructing projects, not necessarily the easiest. For instance, I have eschewed the use of stripboard construction, instead giving only printed circuit board descriptions and layouts. My reason in doing that is because stripboard is, in essence, merely a prototyping method of construction, not the sort of method I want to promote to my readers in the final projects in my book.

My feeling is though, that in a modern electronic circuit, the printed circuit board is as much a component of the final device as any other part. It is not merely a method of making connections between parts of a device — it is a part of the device in its own right! Indeed, in many circuits, how the printed circuit board is made, how it is designed, where all the components are physically placed in relation to each other, and in fact what it is made of, can all define how the circuit works, or even *whether* the circuit works.

Likewise, though not so important to the working of an electronic circuit, the housing used for a circuit is worth bearing in mind. A good housing can help to make a good electronic device; a poor housing inevitably makes a poor device. You can't make a silk purse from a sow's ear, and if your project is housed in a sow's ear of a housing, then no matter how well the circuit works, no matter how much hard graft you've put into it, your project simply can't be a silk purse.

So, on to the projects themselves. There are four, listed with brief details and grading for complexity below.

The four projects:

1 Car-to-USB Power Supply. Allows certain devices (digital audio players, mobile phones and so on) to be powered and recharged in your car. *Easy.*

2 Heart Flasher. Fun device to show your love. *Moderate.*

3 Guitar Headphone Amplifier. Don't wake the neighbours while fingering your frets. *Moderate.*

4 Bench Power Supply. Useful test-bench equipment you can use to power other circuits you build. *Hard.*

Note that even though I class them as being between easy and hard to build, none of them is beyond the scope of most competent DIY-ers and, with the exception of the Bench Power Pupply (which is mains-powered), you shouldn't worry too much about things going wrong, as the only damage you're likely to do even if you bodge up totally is to ruin a couple of cheap electronic components.

Furthermore, if you read this book first, and follow constructional details given here, things shouldn't go wrong at all.

Car-to-USB Power Supply

These days, rechargeable battery-powered consumer devices are becoming increasingly more complex. As many of them are now often linked with computers, they can often be recharged via the computer connection. *Digital audio players* (sometimes called *mp3 players* — a misnomer, as most play music files in other formats than just mp3) are the commonest such device, although some mobile phones and other devices seem to be going the same way.

A computer connection most commonly relies on a *universal serial bus (USB)* connection, from which the device can obtain its recharging power, as the USB interface allows a certain level of power (5 V, at up to 500 mA) through the connector. Away from a computer, however, users have to rely on another source or recharging power if the battery runs low on their devices. Typically, this is in the form of a mains-powered recharger. This is fine if you are within reach of a mains power socket. Not so fine though if you are further afield than that — in the car, say. A useful project if you have a USB-rechargeable device is therefore the Car-to-USB Power Supply — shown in Figure 8.1, and detailed first.

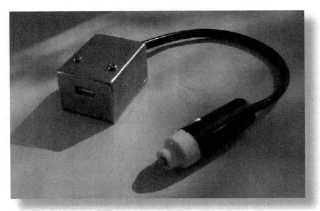

FIGURE 8.1   The Car-to-USB Power Supply.

Principle

The Car-to-USB Power Supply project here uses a car's internal low voltage electrical circuit to allow a USB-rechargeable device to be recharged. It does this by connecting to the car's electrical circuit using a cigar lighter plug that plugs into the car's cigar lighter socket. The low voltage (between about 11 V and 14 V DC) is regulated within the Car-to-USB Power Supply down to exactly 5 V DC, and supplied to a standard USB A-type socket (the same sort that's on any modern personal computer).

The principle of the Car-to-USB Power Supply is shown in Figure 8.2, where these three main stages are identified.

> **Take Note**
>
> The output from the Car-to-USB Power Supply project is not suitable for direct connection to rechargeable batteries, so you shouldn't use it to connect straight to, say, a battery pack. Such packs require some form of control circuitry to ensure only the correct recharge voltage and current is applied. The output of the Car-to-USB Power Supply is a basic 5 V DC regulated supply at up to 1 A, and control circuitry (always present within a rechargeable battery-powered consumer device) is still required.

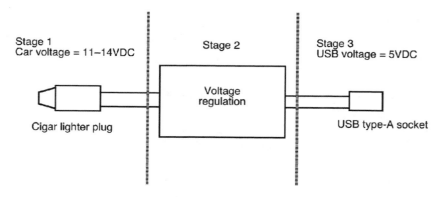

FIGURE 8.2   The Car-to-USB Power Supply principle.

Circuit

The Car-to-USB Power Supply circuit is shown in Figure 8.3. It's not the world's most complicated circuit as you'll appreciate, with the grand total of only five components. It does however, introduce an important principle — that of the integrated circuit voltage regulator (IC1). Voltage regulation is a process whereby an unregulated or varying DC voltage is changed to a regulated or fixed one. In this project a car's main voltage is around 12 V DC, but can vary between about 11 V and 14 V, depending on the state of the car battery and whether the engine is running, so is effectively unregulated. The USB specification, on the other hand, calls for a regulated 5 V DC supply. So what is required is a voltage regulator that takes the unregulated input of 11 to 14 V DC, and gives a regulated output of the required 5 V DC. Integrated circuit IC1 does just that.

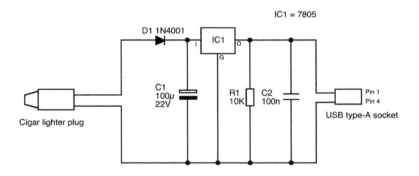

FIGURE 8.3   The Car-to-USB Power Supply circuit.

The remainder of the circuit merely aids this. Diode D1 protects the remainder of the circuit against reverse polarity connection (say, if you wired the project incorrectly so the input voltage was reversed). Capacitor C1 acts as a current reservoir, ensuring that sharp voltage peaks and troughs don't pass through the circuit to the output. Resistor R1 and capacitor C2 aid this.

Construction

Voltage regulator IC1 has a maximum voltage dropped across it of about 9 V (the maximum 14 V car voltage, minus the 5 V output voltage). At the USB specification of a maximum of 500 mA, the maximum power it dissipates is:

$$P = IV = 500 \text{ mA} \times 9 \text{ V} = 4.5 \text{ watts}$$

As a result, there is a potential heat problem if the voltage regulator is free-standing while operating in worst-case conditions, and it requires some form of heatsinking to aid heat dissipation and keep it cooler. The easiest way to provide heatsinking is — and that's the method used here — to bolt it to a metal housing. The metal housing conducts heat away from the voltage regulator quite efficiently.

▷ **FYI** ◁

Actually, voltage regulators like IC1 are protected from overheating by their internal circuitry. If such a voltage regulator gets too hot, an internal thermal cutout control turns the output supply off until it cools down again. Heatsinking merely aids in preventing the cutout situation happening.

None of the components are critical. Any of the 7805 variety of 1 A voltage regulators is usable, and neither the resistor nor the capacitors need to be closely matched. Capacitor C1 should have a voltage specification of at least the car's voltage, so 16 V is the next highest standard limit. The only component which needs matching in any way is the USB socket — as it's a board-mounting variety you should make sure you obtain the correct fitment. Alternatively, you could just adapt the printed circuit board copper foil track layout to suit the one you buy.

The printed circuit board layout is shown in Figure 8.4. Start construction by fitting and soldering the terminal pins, diode D1, resistor R1 and capacitors C1 and C2 into the printed circuit board, checking polarity of the electrolytic capacitor C1 prior to soldering.

USB/car power

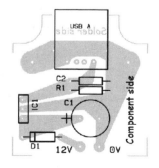

FIGURE 8.4   Printed circuit board layout of the Car-to-USB Power Supply (actual size).

Next up, fit the USB connector to the board. Note that it has four connection terminals, while two extra tabs are used to hold the whole socket in place. The tabs require larger holes (about 2 mm) to be drilled prior to mounting. Bend the lugs of the extra tabs over once the socket is in place on the printed circuit board, to provide extra strength. Solder the tabs to the printed circuit board along with the four terminals.

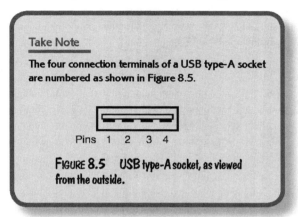

Take Note

The four connection terminals of a USB type-A socket are numbered as shown in Figure 8.5.

Pins   1   2   3   4

FIGURE 8.5   USB type-A socket, as viewed from the outside.

Now fit the voltage regulator IC1 — note though that if you intend using a small diecast metal housing like the one used in the project here, the regulator must be fitted very close to the printed circuit board, with only 1 or 2 mm clearance, in order that there is enough room for the completed printed circuit board to fit inside the housing. To allow the regulator to fit this closely to the printed circuit board, the component holes need to be larger than 1 mm to allow the wide top end of the regulator pins to go through the board — as shown in Figure 8.6. Use a 1.5 mm drill for these holes.

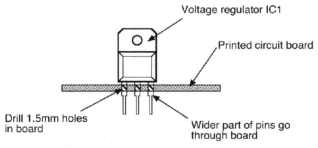

FIGURE 8.6   Fit the voltage regulator close to the board to reduce its height – this way, a smaller housing can be used.

Now solder the ends of the connection lead to the terminal pins. This completes the printed circuit board construction.

Drill the housing to take the cable clamp, the bolt to hold the voltage regulator, and two bolts for spacers. File a slot to take the USB connector. When complete, the housing should look something like that in Figure 8.7.

FIGURE 8.7   Showing holes and slot in the Car-to-USB Power Supply housing.

Fit the two printed circuit board mounting bolts to the inside of the housing, using washers and nuts to act as spacers (to prevent the soldered connections on the underside of the printed circuit from touching the metal housing). Next, feed the free end of the connection lead through the cable clamp hole. Rub some heatsink compound on the voltage regulator tab and on the inside of the housing where the regulator is to be bolted. Offer the complete printed circuit into place inside the housing onto the mounting bolts. Insert the voltage regulator bolt through from the outside, fitting a washer and nut on the inside. Fit washers and nuts to the mounting bolts to hold the printed circuit board in place. Tighten all bolts to hold the printed circuit board firmly.

Take Note

Note that no insulating washer is needed between the voltage regulator and the housing. The metal tab of the voltage regulator is at 0 V, so there is no risk of short circuiting between the voltage regulator and other components, or the housing and other parts within the car.

Finally, connect the cigar lighter plug to the free end of the connection lead. Make sure that polarity is correct — that is, the positive connection is the inner terminal of the plug.

Figure 8.8 shows the printed circuit board copper foil track layout.

FIGURE 8.8   Printed circuit board copper foil track layout of the Car-to-USB Power Supply (actual size).

Parts list

Semiconductors
IC1   7805 voltage regulator
D1   IN4001 diode

Resistors
R1   10k 0.125 watt

Capacitors
C1   220µ, 16 V radial electrolytic
C2   100n polypropylene

Miscellaneous
Printed circuit board
Housing, diecast aluminium (approximate
    dimensions 52 by 38 by 32 mm)
USB type-A, printed circuit board-mounting socket
Car cigar lighter plug
Two-core insulated wire — to take at least 0.5 A
Cable clamp

Heart Flasher

OK, I admit it! This is just a fun thing. It's not a practical project at all really — scoring zero on a usefulness rating, it's just meant for a bit of a laugh. Nevertheless, the project does have some purpose, introducing readers to some solid electronic circuit concepts and providing a handy basis from which to practise some circuit building techniques.

The heart of the project — excuse the pun — is an array of light emitting diodes (LEDs), that flash on then off in an ordered and sequential manner, such that the physical shape of the array forms a heart appearance. Persistance of vision as the LEDs flash on and off means that to the observer the heart shape appears vibrant and pulsing (Figure 8.9). Just the thing to build and give to your loved one as, say a soppy Valentine's Day present, or just for fun.

FIGURE 8.9   Wear your heart on your sleeve with the Heart Flasher.

Principle
The Heart Flasher project has three main sections, shown in Figure 8.10. Section one is formed around a clock circuit that provides a constant electrical pulsing at a rate of approximately 10 cycles a second (10 Hz).

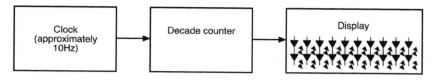

FIGURE 8.10 The principle of the Heart Flasher project.

The next section is a decade counter — a type of circuit that counts the incoming pulses from the clock, and issues output signals to each of ten outputs sequentially. Once the decade counter has counted ten input pulses from the clock, it cycles round to start at the first output again.

The final section of the circuit is an LED array. Some 20 LEDs are used, two per sequential output from the decade counter. The LEDs are formed into the shape of a heart, so that as they are cycled on and off, the visual impact is of a pulsing heart.

Circuit

The circuit of the Heart Flasher is given in Figure 8.11. Integrated circuit IC1 is a 555 timer, connected as an astable multivibrator, to provide a clock signal. Frequency of the astable multivibrator is given by the expression:

$$f = 1.44/C(R1 + 2[RV1 + R2])$$

which means that the frequency can be adjusted between about 50 Hz and 6 Hz.

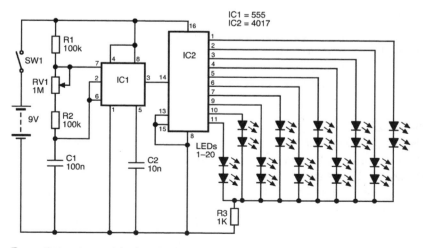

FIGURE 8.11 Circuit of the Heart Flasher project.

The output (from pin 3 of integrated circuit IC1) is taken to the input of integrated circuit IC2, which is a 4017, divide-by-10 Johnson counter with decoded outputs. Integrated circuit IC2 presents a high logic level at each of its decoded outputs sequentially. While the outputs are shown in the circuit as from pins 1–7 and 9–11 of the integrated circuit, the outputs are not sequentially in this order. Figure 8.12 shows how the actual integrated circuit pin numbers correlate to the decade counter outputs.

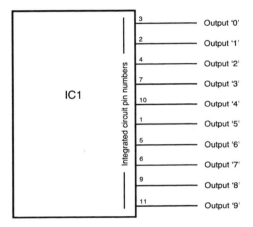

Figure 8.12 Showing how the actual pin numbers of integrated circuit IC1 relate to the decade counter outputs.

Light emitting diodes LEDs 1–20 form the heart-shaped array, driven directly by the decade counter outputs. Two LEDs are driven by each output, which means that the heart shape is actually made from ten pairs of LEDs, rather than twenty individual ones.

Construction

While construction of the project is not too complex, care must still be taken — particularly with the printed circuit board as it features many thin copper tracks and resultant small component pads. Soldering must therefore only be undertaken with a pretty fine soldering iron bit tip, and preferably only using no higher than a 15–20 watt soldering iron. Figure 8.13 shows the printed circuit layout and wiring connection details.

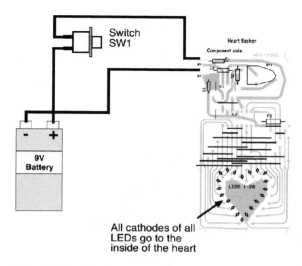

Switch
SW1

Heart flasher
Component side

9V
Battery

All cathodes of all
LEDs go to the
inside of the heart

FIGURE 8.13   Printed circuit board layout and wiring connection details of the Heart Flasher.

First, make sure the printed circuit board fits snugly into the housing, filling out the corners to fit the housing corner posts. Insert and solder two terminal pins into the printed circuit board.

There are some fifteen wire links to be made on the printed circuit board, and these should be inserted and soldered next.

Insert and solder the remainder of the components in the following order: resistors, capacitors, integrated circuits, and finally the twenty LEDs. You can use an integrated circuit holder for integrated circuit IC1, but (if you use the same sized housing as used here) you should not use a holder for integrated circuit IC2 — the 9 V battery lies above this, and if IC2 is in a holder there will not be enough room. Make sure that all polarized components (ie, integrated circuits and LEDs) are the correct way round prior to soldering. Note that the cathodes of all LEDs go to the inside of the heart shape. Also, for visual purposes, make sure that all twenty LEDs are mounted at the same height above the board — approximately 10 mm.

Next, solder the battery connector and switch SW1 onto the printed circuit board.

Finally, mark and drill the housing to mount switch SW1, label the housing to suit, and complete the project.

Press switch SW1 to operate the Heart Flasher, and adjust the flash speed to suit by adjusting preset potentiometer RV1. While the flash speed is calculated to give a reasonable flash rate, you can adjust the overall speed by replacing resistors R1 and R2 with lower values to increase the flash rate, or higher values to slow it down.

Figure 8.14 shows the printed circuit board copper foil track layout.

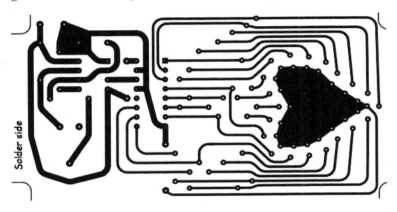

FIGURE 8.14   Printed circuit board copper foil track layout of the Heart Flasher (actual size).

Parts list

Semiconductors
IC1   555 multivibrator
IC2   4017 decade counter
LED1–20 3 mm light emitting diodes

Resistors
R1,2   100k 0.125 watt
R3   1k 0.125 watt
RV1   1M preset potentiometer

Capacitors
C1   100n polyester layer
C2   10n polypropylene

Miscellaneous
SW1   momentary action push button
Printed circuit board
PP3 battery clip
PP3 battery
Housing, transparent plastic (approximate
   dimensions 112 by 63 by 32 mm)
Terminal pins

Guitar Headphone Amplifier

Are the neighbours giving you grief? Are you fed up with the complaints over your racket? Do you wish they'd all go away and leave you to practise alone? Well, with this Guitar Headphone Amplifier your problems could all be solved. You could slash your axe at the volume you want, without disturbing anyone else's peace.

The Guitar Headphone Amplifier here is a neat project that amplifies your electric guitar's output to a suitably loud level for even the most headstrong head basher, but with ordinary earphones. So no one else hears your playing. The grief disappears, and everyone's happy.

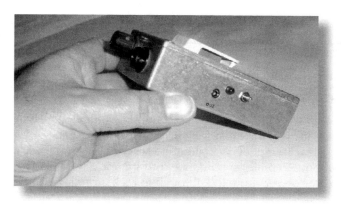

FIGURE 8.15   The Guitar Headphone Amplifier.

Principle

The Guitar Headphone Amplifier is based around two operational amplifiers, which together increase the signal created by an electric guitar when being played in two stages, up to the levels required by a pair of ordinary earphones.

Circuit

The circuit of the Guitar Headphone Amplifier is shown in Figure 8.16, where you can see the two operational amplifiers, which operate in series to amplify the small level signal at the circuit's input up to the levels necessary to drive earphones. The particular operational amplifiers used, while separate electrically, are not actually physically separate devices, and instead are combined in a single integrated circuit IC1. This reduces the size of the circuit, hence allowing a smaller printed circuit board to be used, in turn keeping the physical size of the completed project down to a pocket-sized format.

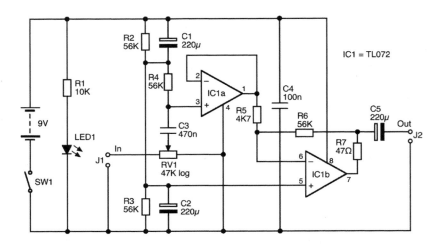

FIGURE 8.16   Circuit of the Guitar Headphone Amplifier.

Potentiometer RV1 allows control of the overall output volume, which in combination with the electric guitar's own volume control, can be configured to give clear sound levels, right up to severe distortion levels. The Guitar Headphone Amplifier should therefore suit most if not all kinds of electric guitar players from folk through to rock, from jazz through to heavy metal.

Operational amplifiers usually require a split-rail or three-level power supply (ie, a power supply with positive, negative, and ground voltages). Operating an operational amplifier from a single supply voltage requires that the voltage (9 V, provided by a PP3-sized battery) be split to give a mid-rail of around 4.5 V, thus emulating a split-rail supply. Capacitors C1 and C2, together with resistors R2 and R3 form a potential divider which does just that.

Construction

While the majority of component holes in the printed circuit board are of the standard 1 mm size, some component holes (for potentiometer RV1, and jack socket J1) should be drilled with a 1.5 mm diameter drill.

Also, if you use the same size of diecast housing as used here, you'll need to file the corner of the printed circuit board for the housing corner post. Also, cut the stem of potentiometer RV1 to the correct length to take its knob (see page 153). These jobs should be carried out before any components are soldered into the printed circuit board.

The printed circuit board layout for the Guitar Headphone Amplifier is shown in Figure 8.17, together with wiring details. Begin construction by inserting and soldering components in the following order: the six terminal pins, resistors R1–7, capacitors C1–5, followed by the larger components jack sockets J1 and J2, and potentiometer RV1.

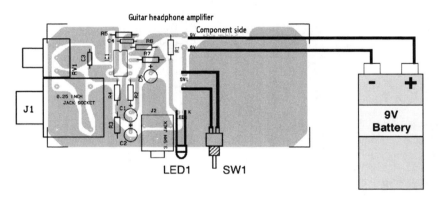

FIGURE 8.17   Guitar Headphone Amplifier printed circuit board layout and wiring diagram.

At this stage you can test the circuit by temporarily soldering the battery clip and switch SW1 to the circuit board. Plug in a guitar (using a standard guitar-style 0.25 inch jack lead) to jack socket J1, and plug in a pair of earphones to jack socket J2.

Once tested, it's time to fit the completed printed circuit board into a housing. Mark and drill holes in the housing for the two jack sockets, the potentiometer, light emitting diode LED1, and switch SW1. Make the holes for jack socket J1 and potentiometer RV1 slightly larger than their actual diameters (by a millimetre or so) then you will be able to wiggle the board into position as the tab of jack socket J2 pushes tightly into its hole in the housing. Using the washers and nuts provided with jack socket J1 and potentiometer RV1, clamp the components to the housing. The printed circuit board should now be held tight within the housing.

Making sure that light emitting diode LED1 is the right way round, push-fit it into the hole you've drilled, and solder its leads to the terminal pins on the printed circuit board. Mount switch SW1 into the housing, and solder its leads to the printed circuit board. Resolder the battery clip to the printed circuit board, and connect the battery.

Finally, mark and drill the holes on the housing lid to mount the belt clip, then bolt the clip to the lid.

FIGURE 8.18   Showing how the printed circuit board fits inside the diecast housing.

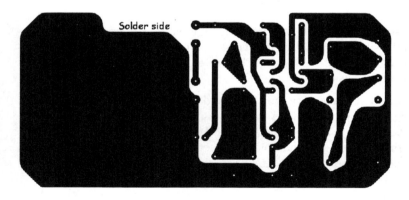

FIGURE 8.19   Printed circuit board copper foil track layout of the Guitar Headphone Amplifier (actual size).

Parts list

Semiconductors
IC1   TL072 dual operational amplifier
LED1   3 or 5 mm light emitting diode

Resistors
R1   10K 0.125W
R2,3,4,6   56K 0.125W
R5   4K7 0.125W
R7   47Ω 0.125W
RV1   47K log miniature printed circuit
   board mounting potentiometer

Capacitors
C1,2,5   220µ 10V radial electrolytic
C3   470n polyester layer
C4   100n polyester layer

Miscellaneous
Printed circuit board
PP3-sized battery
Battery clip
J1 0.25 inch jack socket, printed circuit board mounting
J2 3.5 mm stereo jack socket, printed
   circuit board mounting
SW1   SPST miniature toggle switch
Terminal pins
Housing, diecast aluminium (approximate
   dimensions 110 by 60 by 32 mm)
Belt clip, for housing
Knob (for potentiometer RV1)

Bench Power Supply

If you want to build electronics projects, work with electronic components, experiment with electronic circuits, and do any of the things covered in this book, you'll need some source of electric power. In many cases batteries can be used, of course, and they're fine for low-power circuit requirements, but batteries have several drawbacks. First, they can provide only limited current before their output voltage drops, so any circuit you work with that requires lots of current can't be powered from them. Second, batteries don't last forever and run down (usually when you least want them to!).

A much better source of power is a mains-powered bench power supply (Figure 8.20). These can give significantly greater current and, while connected to mains electricity, always provide the power needed.

To keep the project as simple as possible, the design used here gives a number of fixed voltages, rather than being variable. The voltages directly available are from a dual-rail (ie, positive and negative values) +12 V/0 V/–12 V supply at a maximum output current of about 0.3 A, together with a single-rail +5 V supply up to a maximum output current of about 0.5 A.

FIGURE 8.20   The Bench Power Supply.

These two separate power supplies are internally unconnected though it's possible to connect them externally as you wish, to add to or subtract from each other, this way providing a range of supply voltages.

As most circuits you'll build are powered by only a small number of voltage options anyway, the voltages provided by this power supply should be more than adequate. For example, most operational amplifier circuits require a dual-rail power supply of between about +/– 9 V to +/– 15 V, so the output of the power supply here is ideal. Likewise, much logic circuitry needs a single-rail power supply of between about 5 V and 10 V, so the single-rail power supply here is also ideal.

These standard voltages aside, it's also possible to combine the two separate internal supplies to create a variety of other voltages. For example, by connecting the 0 V terminal of the 5 V circuit to the +12 V of the other circuit, and using the –12 V connection as ground, a 29 V supply can be formed (ie, 12+12+5). Likewise, connecting the 5 V terminal to the +12 V terminal, then using the two 0 V terminals a 7 V supply can be formed (ie, 12–5). Other voltages can be formed similarly.

Principle

A mains power supply has the job of reducing and converting a high mains AC voltage into a low DC voltage. In a standard power supply this is undertaken in four main stages, illustrated in Figure 8.21 together with the waveforms at the various stages.

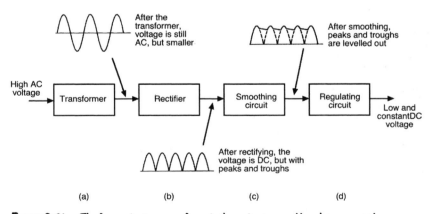

(a)　　　　　(b)　　　　　(c)　　　　　(d)

FIGURE 8.21　The four main stage waveforms in the mains-powered bench power supply.

First, the high mains AC voltage is transformed to a lower AC voltage, as shown at Figure 8.21 (a). After this the AC voltage is converted to a DC one by rectification — Figure 8.21 (b). Next, the voltage is smoothed to reduce the peaks and troughs of the basic rectified voltage — Figure 8.21 (c). Finally, as shown at Figure 8.21 (d), the output voltage is regulated to maintain a constant DC voltage.

The power supply here follows these basic stages, although there are in fact three individual power supply circuits making up the overall design. Two circuits combine together to give positive and negative +12 V/–12 V values, while the third 5 V circuit is unconnected. The main principle of each is the same however.

Circuit

The overall circuit of the power supply is shown in Figure 8.22, where the three individual circuits within the whole circuit can be identified. Transformers T1 and T2 transform 230 V ac mains voltage input to the lower voltages (15 V ac and 9 V ac) required by the rest of the circuit.

Bridge rectifiers BR1 and BR2 rectify the alternating voltages to direct voltages, while capacitors C1–3 smooth these voltages.

Integrated circuits IC1–3 are voltage regulators, which regulate the smoothed voltages into slightly lower, but fixed output voltages.

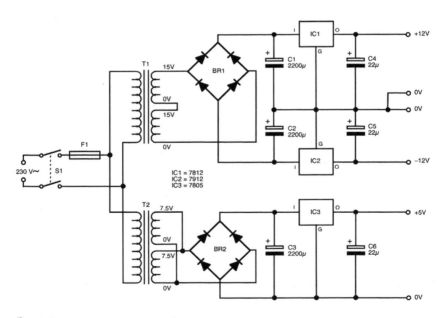

FIGURE 8.22   *Mains-powered bench power supply circuit.*

Construction

The printed circuit board contains most of the components for the Bench Power Supply, and is a compact yet easy-to-build design. Some holes on the printed circuit board should be drilled larger than 1 mm — the mounting holes for the housing and the mounting holes for transformer T1 are all 4 mm diameter, while the terminal holes for transformers T1 and T2 are 1.5 mm. Figure 8.23 shows the layout of the printed circuit board to locate the components.

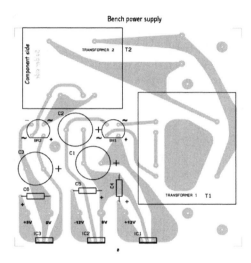

FIGURE 8.23   Layout of the Bench Power Supply printed circuit board.

Now position the printed circuit on the inside base of the housing where it is to be mounted — make sure the positions for the three integrated circuits IC1–3 are against the rear panel of the housing. Mark the holes in the housing base for the three mounting bolts. Remove the printed circuit board, and drill the holes in the housing base, using a 4 mm drill. Insert the 20 mm long M3 bolts and spacers, ready for the printed circuit board.

Insert and solder the seven terminal pins, and integrated circuits IC1–3 into the printed circuit board.

Mount the printed circuit board temporarily onto the three mounting bolts in the housing base. Mark the three integrated circuits IC1–3 mounting holes on the rear panel of the housing. Now remove the printed circuit board again, and drill the holes you just marked, using a 4 mm drill. Make sure the holes have no burrs on their edges, and if burrs exist use a large drill to hand ream the hole edges to remove them.

Now insert and solder all remaining components into the printed circuit board, in the following order: bridge rectifiers BR1 and BR2, capacitors C1–6, transformers T1 and T2. Now, using 6 mm long M3 bolts, bolt transformer T1 to the printed circuit board.

Mark and drill holes in the housing panels for the cable clamp, the panel-mounting fuse holder, and the on/off switch, then mount these components to the housing.

Position the printed circuit board onto its mounting bolts in the housing, but do not fasten it down with nuts and washers yet. Insulating washers must fit between the three integrated circuits IC1–3 and the housing panel, as the three integrated circuit tabs are at different voltages, and bolting them directly to the panel would create short circuits. Each integrated circuit must therefore be bolted to the panel as shown in Figure 8.24, with an insulating washer between the integrated circuit tab and the panel, an insulating bush through the integrated circuit tab hole, and a bolt through the bush. This arrangement ensures that the integrated circuit is electrically isolated from the housing, though rigidly held to it.

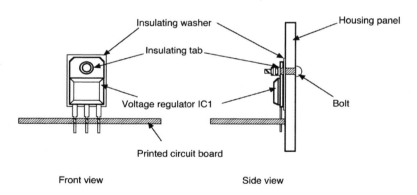

Front view   Side view

FIGURE 8.24   Mounting arrangement for integrated circuits IC1–3 to the housing rear panel.

The panel is also used as a heatsink, dissipating heat safely away from the integrated circuits, so heatsink compound must be applied to all mating surfaces prior to assembly. See page 165 for details of applying heatsink compound to surfaces.

Once heatsink compound is applied, the insulating washers and tabs are fitted, and the three integrated circuits are bolted to the housing rear panel. Complete the mechanical assembly by fastening the printed circuit board with nuts and washers onto the three mounting bolts.

Following the wiring diagram of Figure 8.25, wire up the project. For safety, and following the same sort of layout as in Figure 8.26, keep all mains wiring to one side of the housing, running it from the housing panel-mounted components (cable clamp, fuse holder, and on/off switch) down to the housing base, along the base and up the other panel. Take mains connections to the printed circuit board from the panel-mounted components down to the base, along the base with the other mains

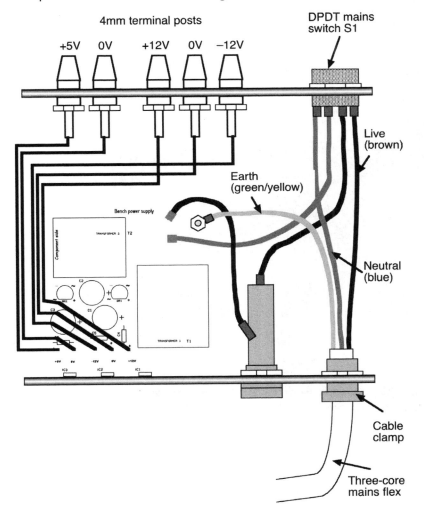

FIGURE 8.25   Wiring diagram for the Bench Power Supply.

wires and directly from the wire group to the printed circuit board. Make doubly sure all mains connections are made tight. Use heatshrink insulated sleeving on all mains connections to make sure that no metal parts remain visible (or more important, make sure they are not touchable).

The five low voltage wires from the printed circuit board to the connectors should be taken to the other edge of the housing to the mains wires, running along the housing base to the front panel, then along the housing base to each connector.

Once wiring is completed, use cable ties or similar to tie the loose wires together into cables, as can be seen in Figure 8.26.

Figure 8.26 Showing the Bench Power Supply innards – note the use of cable ties to form the two groups of wires (mains wiring on the left, low voltage wires on the right).

FIGURE 8.27   Bench Power Supply printed circuit board copper track foil pattern (actual size).

Parts list

Semiconductors
ICI   7812 voltage regulator
IC2   7912 voltage regulator
IC3   7805 voltage regulator
BR1,2   W05 bridge rectifier (1.5A, 50Vac)

Capacitors
C1,2,3   2200µ, 25 V radial electrolytic
C4,5,6   22µ, 16V radial electrolytic

Miscellaneous
Printed circuit board
T1   2 x 15 V, 10 VA printed
   circuit mounting transformer
T2   2 x 7.5 V, 5 VA printed
   circuit mounting transformer
S1   DPDT illuminated push button mains switch
Housing, folded aluminium (approximate
   dimensions 200 by 152 by 78 mm)
5 x 4 mm terminal posts (2 black, 2 red, 1 green)
3 insulating washers and tabs for
   TO-220 style components
M3 (6 mm) bolts, nuts, washers
M3 (20 mm) bolts, nuts, washers
M3 10 mm spacers
FS1   panel-mounting fuse holder
100 mA 20 mm fuse
Length of 1 A mains flex and plug
Cable clamp
Terminal connection pins
Solder tag (M3)

Component suppliers

Over the next few pages is a list of some of the mail order electronic component suppliers in the UK. The list is just a sample of suppliers, and the presence of any supplier on the list is by no means an endorsement. Look in a telephone book for local suppliers. Also, check my website, at:

http://keithbrindley.members.beeb.net

where I'll try to keep an updated list of supplier links and any other relevant information.

2001 Electronic Components Ltd
Eastman Way
Stevenage Business Park
Pin Green
Stevenage
SG1 4SU

Telephone: 01438 742001
URL: http://www.2k1.co.uk

Bowood Electronics Limited
7 Bakewell Road
Baslow
Bakewell
Derbyshire
DE45 1RE

Telephone: 01246 200 222
URL: http://www.bowood-electronics.co.uk

Starting Electronics Construction

Cricklewood Electronics
40-42 Cricklewood Broadway
London
NW2 3ET
Telephone: 020 8452 0161
URL: http://www.cricklewoodelectronics.co.uk

Dannell Electronics Limited
Unit 2 Funston's Commercial Centre
Arkesden Road
Clavering
Saffron Walden
Essex
CB11 4QU
Telephone: 0845 603 1509
URL: http://www.dannell.co.uk

Disponible
Highfield House
College Business Park
Ripon
North Yorkshire
HG4 2SG
URL: http://www.disponible.co.uk

ElectroValue Ltd
Unit 5
Beta Way
Thorpe Industrial Park
Egham
Surrey
TW20 8R
Telephone: 01784 433604
URL: http://www.electrovalue.co.uk

ESR Electronic Components
Station Road
Cullercoats
Tyne & Wear
NE30 4PQ
Telephone: 0191 251 4363
URL: http://www.users.zetnet.co.uk/esr

Farnell InOne,
Canal Road,
Leeds,
LS12 2TU
Telephone: 0871 200200
URL: http://uk.farnell.com

Fast Components Limited
Winchester House
Winchester Road
Walton on Thames
Surrey
KT12 2RH
Telephone: 0870 750 4468
URL: http://www.fastcomponents.co.uk

GreenWeld Limited
14 West Horndon Business Park
West Horndon
Brentwood
Essex
CM13 3XD
Telephone: 01277 811 042
URL: http://www.greenweld.co.uk

Henrys Electronics Ltd
404 Edgware Road
Paddington
London
W2 1ED
Telephone: 020 7258 1831
URL: http://www.henrys.co.uk

JPR Electronics Ltd
4 Circle Business Centre
Blackburn Road
Dunstable
LU5 5DD
Telephone: 01582 470000
URL: http://www.jprelec.co.uk

Magenta Electronics Ltd
135 Hunter Street
Burton-on-Trent
DE14 2ST

Telephone: 01283 565435
URL: http://www.magenta2000.co.uk

Mainline Surplus Sales
Unit 1A Cutters Close Industrial Estate
Cutters Close
Narborough
Leicester
LE19 2FZ

Telephone: 0870 241 0810
URL: http://www.mainlinegroup.co.uk

Maplin Electronics Ltd
National Distribution Centre
Valley Road
Wombwell
Barnsley
South Yorkshire
S73 0BS

Telephone: 0870 4296000
URL: http://www.maplin.co.uk

Mega Electronics
Mega House
Grip Industrial Estate
Linton
Cambridge
CB1 6NR

Telephone: 01223 893900
URL: http://www.megauk.com

Quasar Electronics Ltd
PO Box 6935
Bishops Stortford
CM23 4WP

Telephone: 08702 461 826
URL: http://www.quasarelectronics.com

Rapid Electronics Ltd
Severalls Lane
Colchester
Essex
CO4 5JS

Telephone: 01206 751166
URL: http://www.rapideducation.co.uk

RS Components Ltd
Birchington Road
Corby
Northants
NN17 9R

Telephone: 01536 444222
URL: http://rswww.com

SIR-KIT Electronics
57 Severn Road
Clacton-on-Sea
Essex
CO15 3RB

URL: http://www.geocities.com/sirkituk

Sycom
PO Box 148
Leatherhead
Surrey
KT22 9Y

Telephone: 01372 361421
URL: http://www.sycomcomp.co.uk

Index